Cognitive Behavioral Therapy

A Beginners Guide to CBT with Simple Techniques for Retraining the Brain to Defeat Anxiety, Depression, and Low-Self Esteem

Travis Wells & Seth Goleman

© Copyright 2019 by Pardi Publishing - All rights reserved.

This document is geared towards providing exact and reliable information in regards to the topic and issue covered. The publication is sold with the idea that the publisher is not required to render accounting, officially permitted, or otherwise, qualified services. If advice is necessary, legal or professional, a practiced individual in the profession should be ordered.

From a Declaration of Principles which was accepted and approved equally by a Committee of the American Bar Association and a Committee of Publishers and Associations.

In no way is it legal to reproduce, duplicate, or transmit any part of this document in either electronic means or in printed format. Recording of this publication is strictly prohibited and any storage of this document is not allowed unless with written permission from the publisher. All rights reserved.

The information provided herein is stated to be truthful and consistent, in that any liability, in terms of inattention or otherwise, by any usage or abuse of any policies, processes, or directions contained within is the solitary and utter responsibility of the recipient reader. Under no circumstances will any legal responsibility or blame be held against the publisher for any reparation, damages, or monetary loss due to the information herein, either directly or indirectly.

Respective authors own all copyrights not held by the publisher.

The information herein is offered for informational purposes solely, and is universal as so. The presentation of the information is without contract or any type of guarantee assurance.

The trademarks that are used are without any consent, and the publication of the trademark is without permission or backing by the trademark owner. All trademarks and brands within this book are for clarifying purposes only and are the owned by the owners themselves, not affiliated with this document.

Table of Contents

Introduction 1

Chapter 1: Psychoeducation 5

Chapter 2: Psychotherapy 7

Chapter 3: What is Cognitive Behavioral Therapy? 10

Chapter 4: History of Cognitive Behavioral Therapy 16

Chapter 5: CBT Compared to Other Treatments 19

Chapter 6: Is CBT Right for You? 21

Chapter 7: How and Why CBT Works 24

Chapter 8: CBT Stages, Process, and Common Treatment ... 30

Chapter 9: Essential CBT Techniques and Tools 40

Chapter 10: Understanding Depression 49

Chapter 11: Treating Depression 61

Chapter 12: Fighting Depression with CBT 63

Chapter 13: Understanding Anxiety 74

Chapter 14: CBT for Treating Anxiety 83

Chapter 15: Boosting Self-Esteem 99

Chapter 16: Overcoming Obstacles 108

Chapter 17: Maintaining Positive Mental Health and Preventing Relapse 115

Conclusion 123

Free Mini Guide Reveals: How To **Accelerate Your CBT Results** INSTANTLY With 4 Simple Exercises

As a way of saying thank you for your purchase, I'm offering a Free guide that's exclusive to the readers of this book.

With CBT as your ideal therapy option, you can be certain that you're on the right path towards recovery. However, a huge obstacle that most patients face is they lack the guidance for using CBT exercises to guarantee their effectiveness.

Our **CBT Protocol** mini guide will help you get the most out of this book by providing you with daily CBT activities to help you accelerate your results. You'll discover:

- A 4-step solution you can use in any situation to silence negative thoughts
- One powerful exercise to calm your racing heart during stressful events
- How to turn your favorite hobbies into a medicine to start relieving depression immediately.

To Get Your Free CBT Protocol & Fast-Track Your Results, visit: www.Protocol.MindPerfection.org

Introduction

This book will introduce you to specific cognitive behavioral therapy techniques that can help you overcome your anxiety, depression, or feelings of low self-esteem. It will also teach you a variety of methods for mastering your thoughts and emotions to gain control over your mental illnesses. Anyone who struggles with mental illnesses such as depression, anxiety, panic, worry, anger, phobias, or low self-esteem will benefit from reading this book.

If you've been diagnosed with anxiety or depression, you may find yourself stuck in a cycle of negative thoughts, emotions, and behaviors. You may also feel trapped in feelings of hopelessness, helplessness, and fear, while frequently having thoughts suggesting that nothing will ever get better.

Cognitive behavioral therapy is effective in helping people struggling with these situations. It is based on the idea that our thoughts, feelings, and actions are all connected, and that dysfunctional patterns of behavior are often caused by unhealthy, negative thought patterns. These negative thought patterns, in turn, can have a massive impact on how we feel as well as how we behave.

Consider the following scenario.

Imagine you make a mistake on a project at work, and this is causing you distress. As a result, you begin experiencing negative thoughts about yourself and interpret this situation as you being useless. You start thinking that the reason you made this mistake in the first place is that you always screw things up or you're not very smart. Soon, you begin to realize the outcomes of your error and the consequences you might face at work. So you start to feel anxious about this situation and develop a headache. The next day, you call in sick so that you won't have to face the consequences of your mistake.

This is an example of a negative thought pattern, and as you can see, it is not a healthy response. Cognitive behavioral therapy can break you from negative cycles like these. It introduces you to strategies and techniques that can help you retrain your brain to see things from a positive perspective. By practicing CBT techniques, you gain the ability to see where your negative thoughts come from and decide if there is any actual evidence to sustain them. It also provides you with tools for reshaping negative thinking patterns into positive ones and improves the likelihood that you will react positively in the future.

In considering the previous example, a healthier response would be to see this situation from a realistic lens. It would be beneficial to view this event from a perspective that everyone makes mistakes sometimes, and it's a normal part of life. It would also be more realistic to tell yourself that it's okay to worry about mistakes and that just because you messed up doesn't mean that something severe will happen. Cognitive behavioral therapy can help you learn to respond to situations like these positively.

The goal of this book is to help you understand the possible causes of your anxiety or depression, and how your thoughts, feelings, and behaviors are related.

Cognitive behavioral therapy is a broad concept. The strategies discussed in this book can be adapted and combined depending on the specific challenge or problem you're facing. As you will see, although there may be some overlap, the different techniques presented in this book can help alleviate many types of mental health symptoms. Learning to implement the correct strategies at the right time can have an overall positive impact on many aspects of your life.

Chapter 1: Psychoeducation

A critical first step in overcoming depression or anxiety is to learn as much as you can about it. Psychoeducation is a term that describes learning about a psychological problem, what the symptoms are, the potential causes of the disorder, and other particulars about the illness. If you are suffering from anxiety, depression, or a similar diagnosis, psychoeducation can be a vital first step in treating your symptoms. It can also be an essential part of the recovery process in preventing relapse. Generally, helpful information about mental health disorders can be acquired from your health care provider, on reputable online websites, or from books such as this one.

When dealing with a mental illness, it's helpful to understand that you are not alone in your struggles. Knowing that certain strategies and techniques will reduce your symptoms can go a long way in helping you cope. It can also help you address your issues better and feel more in control. For example, learning about panic attacks can make them less scary when they happen to you, and understanding some of the causes of depression can make you feel less hopeless in your journey to recovery.

As you educate yourself about your diagnosis, it might be helpful to share the information with family and friends so that they can be more understanding and empathetic to your situation. Educating others about your condition can also improve the relationships in your life and further develop your support system. However, everyone is different, and only you can determine if disclosing and talking about your symptoms with others will be beneficial.

Chapter 2: Psychotherapy

Psychotherapy is an umbrella term used to describe a variety of treatments available for mental disorders and diseases. It's a method of therapy that focuses on psychological intervention rather than medical treatment. Psychotherapy is sometimes referred to as 'talk therapy" or counseling and, as a whole, is a popular treatment. The goal of psychotherapy is to help the patient heal in the present and learn effective coping strategies for the future.

Individuals may choose to engage in psychotherapy when they are going through a difficult period, like suffering from loss or feeling overwhelmed. It can be a helpful treatment for those dealing with an immediate situation such as divorce or stress at work. Doctors may recommend psychotherapy as a short-term solution for working through these problems. Psychotherapy is also an effective treatment for people who find themselves experiencing feelings of hopelessness, fear, prolonged sadness, or low self-esteem.

Psychotherapy is generally a collaborative treatment that involves building a positive relationship between the patient and the health care provider. It is practiced regularly by psychiatrists, psychologists, and other mental health professionals across the country. The approach used by psychotherapists creates a neutral, supportive environment that allows patients to talk openly about their struggles. Therapists use a scientifically validated method to help patients work through their problems and learn healthier, more effective behaviors. Commonly, in psychotherapy, the patient and the doctor work together to identify issues that are interfering with the participant's quality of life. The goal is not necessarily to solve the problem, but rather to discover better or new coping skills to deal with the present and future challenges.

There are several approaches to psychotherapy and a variety of theoretical perspectives. Psychologists may combine elements from more than one method, but generally, the goal is the same. The mental health practitioner aims to understand the client and their particular problem to develop and implement both long-term and short-term solutions.

Cognitive behavioral therapy has become known as one of the most effective and routinely practiced methods of psychotherapy. The strategies introduced in CBT can be practiced in collaboration with a trained professional or self-taught and developed without the support of a therapist. This book will provide you with information, strategies, and techniques that can help prepare you to be your own therapist.

Chapter 3: What is Cognitive Behavioral Therapy?

Cognitive behavioral therapy is a short-term, evidence-based psychotherapeutic treatment that is commonly used to treat a range of psychological problems. It has become increasingly popular in recent years. Although CBT is often recommended as a treatment for a wide range of disorders, it was first developed as a management tool for depression. CBT is now considered a very effective treatment for many different disorders, including depression, anxiety, phobias, low self-esteem, problems with alcohol or drug addiction, and anger management issues. CBT is increasingly being prescribed as both an alternative and a supplement to medical intervention. Research has shown that CBT can even reduce symptoms of mental and physical health conditions that some other treatments are unable to relieve.

Key Concepts

Cognitive behavioral therapy is based on the concept that our thoughts, perceptions, and emotions all have a strong influence on our behavior. How we think about a specific situation in our life can directly influence how we deal with it. CBT follows the premise that our thoughts and feelings play a fundamental role in determining our behavior, and that over time, we tend to develop specific patterns of thinking and feeling. If these patterns are destructive, unhealthy, or unrealistic, they can have a negative impact on behavior. Research indicates that the way we perceive a situation may have more of an influence on our reaction to it than the situation itself.

Cognitive behavioral therapy teaches us how to identify and change destructive or disturbing thought patterns that have a negative influence on our emotions and behavior. It aims to improve our negative thought patterns and turn them positive. This is achieved by learning and practicing techniques that empower us to change. By educating ourselves on CBT methodology, we learn how to challenge our distorted thoughts and question whether our beliefs are an accurate depiction of reality. CBT provides us with a new way of understanding our problems and skills to deal with them as they occur. The strategies taught in CBT teaches us to focus on improving our thoughts, mood, and overall functioning for longer-term improvements in mental health and happiness.

Evidence-based

CBT is recognized across the world as an evidence-based therapy, which means that it has been proven to be an effective treatment through rigorous scientific research. It has been evaluated in a scientifically sound way, and findings indicate that it works well for many different kinds of problems. It is currently the only psychological treatment approach with the most scientific support and often recommended for various mental illnesses.

Involves psychoeducation

CBT requires at least some level learning, including education about your particular diagnosis or the specific strategies expected in your treatment. If you have opted to receive CBT from a trained mental health provider, he or she will likely provide you with some information about CBT at the beginning of your treatment. For those who opt to self-learn CBT strategies, there are many helpful sources of information available online. Learning about your illness and the best options for treating it is an essential first step in CBT.

Collaborative

Cognitive behavioral therapy encourages a shared therapeutic relationship between therapists and clients. Generally, the client and therapist will work together to identify and understand the client's difficulties to come up with strategies for addressing them. This requires them to be on the same page, invested in the process, and willing to participate actively. As mentioned,

CBT is a therapeutic approach that can also be done without the involvement of a therapist. To be successful in improving your state of mind, CBT will require your active participation in the methodology and the techniques behind it.

Problem-focused and goal-oriented

Cognitive behavioral therapy takes a hands-on, practical approach to problem-solving. It's a goal-oriented approach that focuses on specific, present-day challenges, as well as uncovering solutions to these challenges. The methods involved in CBT encourage education and skill development. They focus on understanding thoughts, behaviors, and feelings as a critical element of therapy. The techniques introduced in CBT are directed towards solving current life problems and teaching long-term solutions.

Short-term

If you are working with a trained therapist, you will find that your CBT treatment will most likely last anywhere from five to twenty sessions. This therapy is meant to be a brief, time-limited service. The length of treatment can vary depending on the severity and complexity of your problems. CBT aims to help clients meet their goals quickly and teach them skills that they can benefit from later. Essentially, therapists in this practice want to work themselves out of a long-term role to leave treatment in the hands of the client. Reliance on the therapist is not encouraged in CBT, as the focus is on educating and empowering the client.

Structured

Sessions with a cognitive behavior therapist have a specific structure and focused approach. The therapist often takes on an instructional role, working with the client to create a plan for the sessions. Together, they will make sure that they cover what will be most beneficial to the client. Generally, there isn't a great deal of free-flowing talk or delving into the past. Clients are directed to discuss specific problems and concerns.

This structure may be replicated for those who wish to practice CBT strategies on their own. The key is to set realistic and specific goals to achieve and to stay focused on them. It may be best to set aside a particular time each day to work on the strategies introduced, and it will be imperative that you commit to the therapy as you would with professional treatment.

Variations available

Cognitive behavioral therapy can be adapted and tailored to suit a wide variety of needs and preferences. The strategies, process, and protocols recommended in CBT can be modified and combined as needed.

Although CBT is often implemented as one-to-one therapy, there are several alternatives that can be quite effective and convenient. CBT may be offered in group settings for those who would benefit from sharing their experiences with others. Generally, group sessions are created for individuals who are suffering from a similar diagnosis or illness. Both individual and group sessions may be facilitated by a peer or a coach rather than by a trained professional. Additionally, information and teachings can be delivered in a variety of different formats. For example, sessions may be available online or over the phone with a therapist rather than in person. Many CBT self-help materials are also available in books, apps, or online, and can be used on their own or as a supplement to professional treatment.

Client becomes therapist

The strategies and techniques outlined in cognitive behavioral therapy are skills you can practice on your own without the intervention of a therapist. This is an element of therapy that is unique to CBT. By educating yourself, following the processes, doing the practice exercises, and analyzing your responses, you can learn new ways of coping that you can continue to benefit from for a long time.

Chapter 4: History of Cognitive Behavioral Therapy

Although cognitive behavioral therapy is often thought of as a modern form of treatment, it is not a new therapy by any means. The fact is that CBT has quite a long history and was developed based on decades of scientific research. It arose from two well-known, distinct schools of thought: behaviorism and cognitive therapy.

Behavioral therapy – Behavioral therapy for depression and anxiety first emerged in the 1950s. It was based on the idea that behaviors can be observed, measured, and modified, and that our responses to stimuli around us shapes our behavior.

Cognitive therapy – Cognitive therapy was developed as a treatment in the 1960s. The idea behind cognitive therapy is that how a person thinks about their experience will have a significant impact on how they feel and how they react to it. Emotional distress arises from thoughts about a situation, more so than the actual event itself.

In the 1960s, during his practice as a psychiatrist, Doctor Aaron Beck began to concentrate on the idea that the link between thoughts and feelings was a crucial one. He found that depressed patients often experienced emotion-filled thoughts that seemed to arise spontaneously. Often, they were not fully aware of these thoughts. Beck turned his attention to what he called "automatic thoughts."

Beck developed this theory further by concluding that, if a person was feeling upset in some way, his or her thoughts were usually negative or unrealistic. He began working with his patients to help them identify and understand their automatic negative thoughts. In doing so, he found that patients were able to think more realistically, understand their problems better, and start overcoming their difficulties. Beck learned that identifying negative thoughts was the key to long-lasting, positive change. His patients felt better and functioned well as a result. By changing their beliefs about their situation, they changed their present behavior and future actions.

Beck called this model "cognitive therapy" because of the focus on thoughts and their role in mental health. When combined with behavioral techniques for the treatment of depression, this new approach was coined "cognitive behavioral therapy," and in a short time, this model of treatment started gaining acceptance in the field of psychology. Since the 1960s, CBT has undergone successful scientific trials in practices around the world and has been applied with success to a variety of problems.

Chapter 5: CBT Compared to Other Treatments

Cognitive behavioral therapy is one of the most widely researched types of psychotherapy, and it differs from other types of treatment in a few notable ways. For one, CBT is generally very problem-focused, structured, and practical, making it easy enough to measure the outcomes and results of this therapy. However, the number of sessions provided by a therapist varies, depending on the individual's needs.

The goal of CBT is to help patients determine their most pressing problems and come up with solutions and strategies to solve them. Unlike some types of talk therapy, CBT provides for a clear service plan with proactive strategies for preventing relapse. When following the processes outlined in CBT, sessions are quite structured, and clear goals can be determined. The objective is to tackle these specific problems one by one while making long-term improvements to your overall well-being.

The nature of the relationship between the cognitive behavior therapist and the client can be different from some other treatment approaches. The therapist will work together with the client to find solutions rather than just telling the client what to do. CBT offers a more collaborative, equal relationship and does not encourage dependence on the therapist like other treatments can. As mentioned, the goal is to have the client learn to do the strategies on their own in the long-term, without the need for a therapist.

CBT offers treatment that is individualized as it is applied to the client's specific situation and problem. Intervention is made effective by choosing strategies and skills that are most appropriate and fitting to the client's needs. The methodology might be a little different from other treatments, involving activities such as role-playing, guided discovery, and journaling exercises. The strategies recommended in CBT are useful and practical; they can be used long-term and implemented without the ongoing involvement of a therapist.

CBT involves homework and self-analysis. It is an active and educational approach, with the client playing a significant role in recovery. When engaged in CBT, the client will monitor their thoughts and feelings, and write them down in logs. This treatment attempts to address specific issues and does not concentrate on the patient's past as many types of therapy do.

Chapter 6: Is CBT Right for You?

Cognitive behavioral therapy was initially developed as a treatment for depression. Research indicates, however, that in recent decades, CBT has become an effective treatment for a wide range of other emotional problems. At this time, CBT has been shown to help with managing bipolar disorder, generalized anxiety disorder, phobias and panic attacks, as well as schizophrenia and other serious mental illness.

CBT can also help manage long-term issues such as addictions, obsessive-compulsive disorder, post-traumatic stress disorder, and eating and sleep disorders. It can be an effective treatment for anger management issues, low self-esteem, and long-term medical health issues such as irritable bowel syndrome and chronic fatigue syndrome. In the case of physical ailments, CBT will not help cure the disorder but can help people learn strategies for how to cope with their symptoms.

CBT can be very effective at helping people deal with emotional challenges or short-term obstacles that they may be facing. It can help individuals work through specific goals, problems, and troubling symptoms during times of stress or conflict. The strategies introduced in CBT can also help improve an individual's social and communication skills, help them cope with grief and loss, or overcome emotional trauma such as abuse or neglect. CBT can have a positive impact on personal relationship and self-confidence, and improve an individual's skills in relating to others. It can also help control mood swings and enhance sleep quality.

Cognitive behavioral therapy can be a helpful treatment for people of all ages, even children. The type of person who might benefit most from CBT will be those who believe in it and appreciate it's methodology. To be effective, the individual must be ready and willing to put in time and effort. They will need to commit to the therapy, put in the work, and agree to do the self-assignments and homework regularly.

With ongoing practice, CBT strategies can be an excellent tool to have available during difficult times. The expectation is not that CBT will solve all problems and remove all symptoms. It may only make the symptoms that people are experiencing less debilitating. CBT is meant to be a short-term treatment — it is about teaching people how to cope with the symptoms. Because CBT helps people develop better long-term coping skills, it can be a far more effective treatment than medication alone. Additionally, research indicates that people experiencing anxiety and depression are less likely to relapse when treated with CBT. Given these statements, CBT is having a very positive impact on the field of psychotherapy.

Chapter 7: How and Why CBT Works

Exactly how cognitive behavioral therapy works to improve mental health is a complicated question. It does seem apparent though that for CBT to be successful, it may help to combine different strategies and techniques individually. We will discuss many of the specific techniques recommended by CBT later.

There are also several core concepts of CBT that may contribute to this type of treatment having lasting, positive results. They are as follows.

Changing Core Beliefs

Having negative thoughts about an event or situation can make it difficult for us to deal with it, making it far worse than it really is. Negative thoughts can lead to poor reactions, which may be displayed by troublesome behavior and, in some cases, psychological problems. Negative thoughts and feelings can trap us in a vicious cycle. It's critical that we understand the influence that cognition has on emotions.

When a person has anxiety or depression, their perceptions and interpretation of events can get distorted. This distorted view of reality can lead to an even more negative mindset, causing the cycle of depression to continue. The primary goal of CBT is to change these negative patterns of thinking to influence behavior more positively.

Changing Behaviors

CBT teaches methods for breaking down problems and situations into smaller parts and learning more effective ways to handle them. Exercises, strategies, and skills are introduced to help you deal with overwhelming problems better. New forms of coping can lead to positive change in attitudes and behaviors.

Those with anxiety and obsessive-compulsive disorder can learn how to integrate fearful things back into their life. Depressed individuals can learn more positive ways of interacting in their environment. These new, positive habits can alleviate many of the symptoms of mental illness, allowing them to live better, healthier lives.

Learning Coping Skills

The goal of CBT is to teach new skills and strategies for dealing with problems. At the core of learning new skills, there will self-monitoring, especially for those who will be completing CBT activities without the support of a therapist. Self-monitoring involves examining, recording, and analyzing your issues so that you can learn and address the symptoms that are affecting you. You will learn how to do this effectively through CBT while developing strategies to help you cope with similar situations in the future.

By confronting fears and problems gradually, you can gain confidence and strength in your ability to cope. Little changes can go a long way towards dealing with big issues. Therefore, when practicing CBT, it's imperative that you feel safe and confident in trying new strategies, as your improvements and growth will help shape the way to your progress.

Solving Life Problems

The techniques introduced in CBT can be useful in helping you solve problems that have become an issue in your life. We sometimes get stuck in situations and environments that we don't know how to get out of, like an unfulfilling job or an unhappy marriage. CBT can teach you new approaches for dealing with the feeling of being stuck and other problems. The techniques introduced can show you how to define your problems clearly, brainstorm solutions, and take proactive steps to recovery. You will learn how to take the initiative to reach your goals systematically. By discovering strategies and techniques that work for you, you can practice developing these skills and learn how to apply them to other similar problems that you face. CBT is about reducing symptoms, empowering you as an individual, and helping you gain confidence for the future.

Retraining Your Brain

Cognitive behavioral therapy emphasizes the importance of identifying, challenging, and changing how we view situations and circumstances in our lives. We know that our patterns of thinking can cause us to see the world around us differently. CBT makes us more aware of how these thought patterns influence both our emotions and actions.

Our brains are responsible for our ability to think, reason, remember, and feel. The information captured in our minds leads us to behave and act in particular ways. The fact is that our brain can get used to seeing things a certain way and spring into action automatically. We may automatically think negative thoughts without even realizing it. If we are trapped in a cycle of negative thinking, our actions and behavior will likely reflect this.

The good news is that we can retrain our brains to function healthily. The brain is like a muscle, and if you exercise it regularly, it can also grow stronger and more efficient. We can learn why we respond to certain things the way we do and how to react to them differently in the future. We can also overcome some of our difficulties when we understand them better.

By using CBT to change the way you think, you can influence your natural, automatic responses to situations in the future. You can impact the part of your brain that holds rational thoughts and responses. Training your mind to be more positive can be accomplished by using strategies such as purposefully remembering positive experiences and good times in the past, and concentrating on positive elements in your life. We will talk more about this later.

As you acquire better thought processes, it may become easier and more natural for you to solve your problems constructively. This will help you feel more confident and in control. Both of these emotions are good for your mental and physical health.

Calming Your Fear Centers

Many of the techniques described in CBT are designed to calm the fear centers in your brain. When you are faced with a situation that causes you distress or anxiety, you will generally have a physical reaction to this fear. Exercises such as deep breathing and other relaxation exercises can help naturally lower your stress levels and help you feel more calm and relaxed. When you are relaxed, your mind will work more efficiently, and you're more likely to find a realistic solution to your stress. When you learn how to relax during times of stress, it will help you to process information better, and the overactive fear response that you are having will return to normal.

Chapter 8: CBT Stages, Process, and Common Treatment

If you are working in collaboration with a therapist, your CBT sessions will generally last between thirty and sixty minutes. As an introduction to the therapy, your provider will likely share some information about CBT and discuss whether or not CBT is the right treatment for you. He or she will then probably ask you questions about your past to help you decide what you need to work on first.

Cognitive behavioral therapy is often described as guided self-help. With or without the collaboration of a therapist, you can use CBT to gain some useful insights and skills to work through your problems. If you are interested in practicing CBT without the support of a therapist, many of the same processes and stages can be completed on your own. As an option, there are some interactive tools available online and in this book.

With some background information on CBT and some practice, you can accomplish the following by acting as your own therapist:

- Learn to identify your problems and issues more concretely
- Become more aware of your mood and emotions
- Develop an understanding of negative automatic thoughts
- Learn to challenge the assumptions that you make in your mind
- Start to distinguish between what is fact and what is just a thought
- Make sense of overwhelming problems by breaking them down into smaller, more manageable parts
- Begin to look at situations from a different and more positive perspective
- Learn strategies for facing your fears and anxieties
- Stop hiding behind avoidance techniques
- Learn to avoid cognitive distortions, generalizations, and "black and white" ways of thinking
- Stop being so hard on yourself and taking the blame for things that are not your fault nor your responsibility
- Stop focusing on how you think things should be and learn to appreciate how they actually are
- Set and achieve goals for better, long-term mental health

Keep in mind that cognitive behavioral therapy is most effective as a short-term treatment centered on helping you deal with specific problems. As you get started, you will want to analyze what your issues and needs are. Think about the problems that you are currently dealing with in your life and where you would like to focus your attention. You might want to begin by asking yourself some of these questions.

- Is your depression or anxiety interfering with family, work, and social life?
- What events are related to your problems?
- What would you like to achieve through therapy?
- Which of your symptoms are most troublesome?
- How are you sleeping? Are you dealing with insomnia or disrupted sleep?
- Are you experiencing issues with personal relationships?
- Are you in an unhappy marriage?
- Are you having difficulty focusing or concentrating at work?

The following section includes information about the stages and methodology involved with cognitive behavioral therapy. It describes what you can expect as you work through the process and how to be successful in doing so. Many of these elements will also be discussed in further detail later.

Assessment Phase

CBT is based on the idea that we often experience emotions that can lead to faulty beliefs and problematic behaviors. These behaviors can have a negative impact on our social life, relationships, work life, the activities that we once enjoyed, and our eating and sleeping habits. When we are distressed or upset, our view of a situation may not be very realistic.

Therefore, changing the way we think can help us change our reactions in specific situations.

During the assessment phase of CBT, your goal is to draw up a list of the problems that you want to competently and effectively master as a result of treatment. You will identify, examine, and break your problems down into smaller parts. This can be accomplished by keeping a diary, journal, or a thought log, recording the different emotions, thoughts, and feelings that you experience in a given period. Use the journal to record different life events and your reactions to them. It might help to make very clear distinctions and break your activities down into the following categories: thoughts, emotions, physical feelings, actions, and situations. By keeping a journal of any incidents that provoke feelings of anxiety or depression, you can begin to identify the thoughts and circumstances within your life that may be contributing to your problems.

Questions to ask yourself during the assessment period include:
- What thoughts and images went through your mind?
- Which situations bothered you in your day?
- What did your various thoughts mean to you? How did they impact you?
- What emotions did you feel? How would you describe them?
- What did you notice in your body? How did the physical sensations feel?

- How did you feel after the event?
- Were you left tired and lethargic, or agitated and hyper-alert?

Practice noticing, recording, and describing your thoughts, feelings, and actions regularly. The more you do this, the easier this strategy will become for you.

Define Problems and Set Goals

During the process of defining your problems and setting goals, you will begin to delve deeper into understanding how your thoughts, feelings, and actions are related. In this stage, your goal is to learn how to examine and evaluate your thoughts during problematic situations. It is during this step that you will attempt to identify and break down your thought patterns and reactions. As you become aware of your thoughts and the impact that they can have on your day to day life, you will be in a better position to do something about them.

This is where the "cognitive" piece of the therapy is focused. You will learn to question the validity of your thoughts and emotions, decide if they are unrealistic and problematic, and set the tone for exploring healthier, more realistic, and more appropriate ways of thinking. You will essentially uncover some of your core negative beliefs and learn how they may be affecting you.

To do this, you will want to review your journal to look for patterns and determine how your thoughts and feelings affect each other. Reflect on your journal and identify which situations impacted your emotions the most and caused you the greatest distress. Look for links between your thoughts and behaviors. How did you react to each situation as it occurred? Look for patterns of negative thought and distorted perception. Use self-evaluation to reflect on your journal accurately. Are your thoughts and feelings often unhelpful and unrealistic? Did you jump to these thoughts without considering all parts of the equation?

Make sure that you look for specific negative thought processes, such as overgeneralizations, all or nothing thinking, and patterns of rejecting the positives and focusing on the negatives. Analyze whether or not your thoughts are realistic given the circumstance that you were faced with, or if they were catastrophized and blown out of proportion.

Practice assessing your reactions and behavior to different events, and this will get easier with time. The next step will be to come up with ways to prove your unhelpful assumptions and beliefs wrong.

Focus on Your Behaviors

During this stage of therapy, you will focus on the actual behaviors that are contributing to your problems. You will assess and analyze which parts of your daily routine and activities have been impacting your mood the most. You will also determine what behaviors you would like to change to improve your life in the long-term.

Consider the behaviors that you recorded in your journal or thought log during the assessment phase. Ask yourself the following types of questions:

- What did you do to get through the situation?
- Did you use any strategies or techniques to help you cope?
- What did you avoid doing?
- What were the automatic physical reactions that you had?
- How do you think your behavior looked to other people?
- What happened immediately after the situation?
- How long did it take you to return to your normal state?
- What could you have done differently?
- How would someone else have handled this situation?
- What will the consequences of your behavior be?

The focus during this stage is not only to analyze your recent behaviors but also to explore healthier behaviors you should express in the future. For example, over time, you may have developed certain behavior patterns that you hide behind to experience less stress and anxiety. These behaviors may be helpful in the short term, but not so much in the long run.

You may feel anxious in social situations and thus avoid these situations altogether to avoid feeling anxiety. Over time, this can lead to increasing social isolation and increased anxiety. During the process of focusing on your behaviors, you may be able to come up with better responses (rather than avoidance) to situations that cause you anxiety.

You will want to think about the goals that you would like to accomplish and the skills that you would like to gain. It is probably a good idea for you to select a few simple, straightforward goals at this time that will be easy to plan for and obtain. Create an outline or a plan for working on these skills in your natural environment.

Confronting fears and anxieties can be very difficult, especially if introspection is something new to you. Make sure that you don't jump in too quickly and take time to reflect as you need it. Generally speaking, cognitive behavioral therapy is a gradual process that can help you take incremental steps towards reaching healthier behavior. Soon enough, you will be ready to practice doing tasks that will help you challenge and get over your irrational beliefs.

Practice the Strategies

To be successful with your therapy, it is imperative that you practice the strategies that will help you cope with everyday life. This is often called having a "plan of positive action." Your plan may include a list of activities, graduating from easy to more difficult to achieve. Give yourself homework assignments to work on these activities one at a time. Be sure to analyze your results; this is where the learning and the skill building will take place. Some of your homework activities may include practicing scenarios, learning how to question upsetting thoughts, and replacing them with more healthy ones. In these exercises, you may be involved in gradual exposure to things that cause you anxiety and fear, journaling activities, analyzing your thoughts and feelings, and practicing ways to calm the body and mind.

Try these exercises while you are feeling comfortable at home before trying them out in more stressful situations. This will help you manage your fear and anxiety while preventing your symptoms from getting out of control.

Record Your Progress

Keep track of which strategies are working for you and which have not been as successful. If something is not working, look for alternatives. What else can you try? What may have worked for you in the past? Is there something in particular that gets in the way of a technique working for you? Pay attention to how you learn best and which strategies and techniques are the most effective.

Continue to Practice

It is important that you continue to practice and develop your skills on a regular basis. What to do, how to do it, and when to do it are all important concepts in CBT. If you are in the right frame of mind and ready for change, you may see the results of your practice immediately. As you start to feel empowered and your symptoms reside, you will want to challenge yourself to go further. Finding some balance will be important. Make sure that you are ready for change but do not push too far, too fast. Recognize healthy situations from unhealthy ones and avoid doing things that will make you feel worse. Once you start realizing some success, continue to apply the principles of CBT to your daily life.

Chapter 9: Essential CBT Techniques and Tools

CBT uses a variety of cognitive and behavioral techniques, and for your treatment plan, you may consider combining any number of them. A lot of the tools introduced in this type of therapy may help deal with everyday situations. This section of the book will outline fifteen different strategies that are common in CBT. Keep in mind that this is not an exhaustive list by any means. Additionally, we will revisit some of these techniques later as we delve further into CBT treatment for depression, anxiety, and low self-esteem.

There are many worksheets you can find online and download for free during this phase of the CBT process, many of which are intended to support you through your growth. Consider looking for templates and worksheets on thought logs, rating scales, opinion checklists, and fear hierarchies.

Journaling

Journaling is a way of keeping a record of your thoughts, moods, emotions, and behaviors, as well as what led you to these experiences. It is a way of gathering data about your thoughts and feelings. When journaling, it is important to make a note of the source your emotion or thought, and the intensity of it, as well as the environment or situation that you were in at the time. The point of journaling is to help you identify maladaptive thought patterns and to understand the impact that these can have on behavior. The ultimate goal, of course, is to learn how to change or adapt thought patterns to be more positive. In CBT, journals are sometimes referred to as "thought records." We will discuss journaling, as it is applied to treat depression and anxiety in upcoming chapters.

Unraveling Cognitive Distortions

Most of us carry around our own collection of "cognitive distortions." Cognitive distortions are essentially faulty ways of thinking. They are inaccurate thoughts in our head that tend to reinforce negative thought patterns and supply us with a false reality. Cognitive distortions include bad habits such as black and white thinking, jumping to conclusions, filtering positive thoughts, focusing on the negative, using overgeneralizations, and catastrophizing situations. It is an essential skill in CBT to discover what your personal cognitive distortions are. Using the techniques of CBT, you will figure out which of these you fall into most often so that you can learn how to stop doing them.

Cognitive Restructuring

Once you have identified some of the automatic thoughts and inaccurate views on which you have been relying, you can begin to challenge them. Explore how these distortions came about and why you still believe them. Also, think about the advantages and disadvantages of having the views that you have. For example, you might believe that to be considered successful in life, you need to have a high paying job. If you don't, you may get discouraged and depressed. Instead of accepting this faulty belief that causes you to feel sad, you could try some cognitive restructuring. Use the opportunity to think about what being a successful person really means to you.

Exposure Therapy

Exposure therapy is a technique that is used most frequently for those who suffer from obsessive-compulsive disorder (OCD), panic attacks, and phobias. You can practice this technique by exposing yourself to whatever it is that makes you anxious or afraid, a little bit at a time. Generally, you will learn some relaxation techniques first, and you will do your best to keep symptoms under control during the limited exposure. Journaling is sometimes combined with exposure therapy so that you can record and understand how you felt during the exercise, and how you managed the negative feelings you encountered. Exposure therapy can take place in a controlled setting like a clinic, in your home, or out in the community. We will look at exposure therapy in more detail later when we examine CBT and anxiety disorder.

Interoceptive Exposure

Interoceptive exposure is another technique used to treat panic disorder and anxiety. It involves exposure to feared bodily sensations to elicit a response. The purpose is to challenge the unhealthy and automatic thoughts that are associated with these sensations and manage them in a controlled environment. During interoceptive exposure, individuals learn to maintain the sensation without panicking to understand that the feelings they are experiencing are not dangerous or life-threatening.

Play the Script Until the End

This strategy involves examining what the worst case scenario is in a given situation. It is especially helpful for those dealing with intense fear and anxiety. This exercise is beneficial in helping you determine what your underlying fear outcomes are. The idea behind this technique is to conduct a thought experiment or a "rehearsal" in your mind. You set out to imagine the worst possible outcome for a situation and then let the event play out in your mind. By doing this, you can learn that no matter what happens, things will likely turn out okay.

Deep Breathing

Although not unique to CBT, calm breathing is another strategy that can be used to relax the body and mind. Deep breathing can be a great tool for calming your nerves during a stressful event. To do this, breathe in through the nose, take a pause for 2-3 seconds, then breathe slowly back out through the mouth. Repeat this for several minutes. Focusing on your breath helps to regulate and slow your breathing, bringing you calmness and peace. It's also useful for clearing your mind to help you to think more rationally.

Progressive Muscle Relaxation (PMR)

Progressive muscle relaxation is a very well-known technique both inside and outside of cognitive behavioral therapy. It involves the repetitive tensing and relaxing of different muscle groups, one at a time. Generally, at the end of PMR, you are left in a more relaxed state of body and mind. PMR and deep breathing will both be a focus when we visit the section on anxiety and CBT.

Behavioral Experiments

In CBT, behavioral experiments can be put in place to test the validity of your thoughts and beliefs. If you believe something to be true, you can set out to test whether or not it is by performing a behavioral experiment. This process involves setting up a test situation and monitoring the results. You can design tests to collect information that may either prove or disprove your beliefs, basically testing the hypothesis that your thoughts are inaccurate.

You might, for example, use a survey to gather information from others or perform a specific action on purpose to see what the results are. The results may not match what you presumed. For instance, if you have trouble sharing your opinion around unfamiliar people, you could try to share your opinion once when you are in a fairly comfortable situation, and observe the outcome. Did you upset anyone involved? Did they dismiss what you had to say? Did it result in the person liking you less? Are you making any assumptions about this? You can try different approaches on different occasions and see what the results are.

Situation Exposure Hierarchies

Situation exposure is a technique that is often recommended for those with OCD. The idea is to be around the situation, event, or object that normally causes anxiety and compulsive behavior, and refrain from engaging in that behavior. This strategy is often paired with both journaling and relaxation techniques. When practicing this technique, you would write down everything that you generally avoid on a list. The next step would be to rank the items in order from lowest to highest impact, essentially creating a hierarchy. You will start with the easiest item on your list and eventually make your way through all of the items as you manage to keep your anxiety under control.

Imagery-Based Exposure

Similar to situation exposure, imagery-based exposure is another strategy in CBT that can be helpful for those dealing with OCD and anxiety. Instead of being directly exposed to a situation or event that causes negative feelings, in imagery-based exposure, you only need to imagine or remember an event that caused these emotions. While bringing to memory the recent negative event, you are directed to remember the sensory details, the emotions that coincided with the event, and the behavioral responses that you experienced. The expectation is for you to continue visualizing these details and practice your relaxation strategies until your anxiety level reduces.

Cognitive Rehearsal

Cognitive rehearsal is a strategy in which you recall a problematic situation or event from your past and work on finding a solution for it. You will concentrate on the details of the event and the negative thought patterns and behaviors that went along with it. You will then begin to rehearse positive thoughts in your mind, thinking about things that you might be able to do differently next time. You'll also look for solutions to this problem in the future, which will make positive changes to your thought processes should the event happen again. This technique sets you up for future success.

Validity Testing

Validity testing is a technique used to help us challenge our ingrained thoughts and beliefs. It involves creating a list of examples that will support or validate the thoughts that we believe to be true. We create the list to defend our viewpoint and prove that our automatic thoughts are correct. In actuality, this technique can help expose the falsehood of our ideas. When we cannot find proof or evidence to validate what we think and feel, we may begin to doubt the authenticity of our thoughts. In this way, negative thoughts may be replaced with realistic, positive thoughts, and we can start to normalize our thoughts, feelings, and emotions.

Activity Scheduling

Activity scheduling is an effective tool for easing symptoms of depression. It is an exercise that helps people engage in behaviors that they have otherwise been avoiding or not participating in as frequently as they used to. The first step is to identify several behaviors or activities that are rewarding and are not happening as often as they should be. Next is to create a schedule of dates and times throughout the week to engage in the activity. You might set the goal of trying to plan to do one pleasant activity every day. It does not have to be too complicated or time-consuming, just something that makes you feel good even for a little while. Doing activities that produce pleasure and positive emotions in your daily life will help make your thinking less negative.

Guided Discovery

The guided discovery technique is another way to help you learn about and understand your cognitive distortions. It is generally practiced in collaboration with a trained therapist, as the therapist can best assist and guide you to understand your thought processes. The therapist will ask you questions about your thoughts, feelings, and behaviors to get to the root of the problem. The point of guided discovery is to help you learn how to alter the way you interpret and process information and change how you look at the world.

This list of strategies and techniques is far from exhaustive but will give you a good idea of some of the tools that are commonly used in cognitive behavioral therapy.

Chapter 10: Understanding Depression

Most people feel sad or unhappy at some point in their lives; it can be a perfectly normal reaction to some of life's difficult situations. However, when intense sadness lasts for a prolonged period and interferes with normal activities, it may be something a little more serious, such as clinical depression. Depression is the most common mood disorder, affecting millions of people in this country at any time. Depression can have severe symptoms and can impact how you think, feel, and manage daily activities. It generally involves feelings of unrelenting sadness, a sense of loss, and includes feelings of both helplessness and hopelessness. Doctors may diagnose depression when symptoms persist for at least two weeks or longer. If someone receives a diagnosis of depression, this means that they are not just going through a temporary bout of sadness or a change in mood; it means that they have a severe mental illness.

There are many emotional, physical, cognitive, and behavioral symptoms of depression. Some of these symptoms are as follows.

Emotional symptoms

A person experiencing depression may feel despair, intense sadness, hopelessness, and have little interest in doing things that they used to enjoy. They may easily feel overwhelmed, and actively avoid social situations and interactions. They can also feel an unrelenting desire to cry throughout the day. The emotional symptoms of depression may be combined with increased anxiety and feelings of anger, hostility, and aggression. There may be episodes of self-loathing and feelings of worthlessness or guilt. Those suffering from depression often feel irritable and restless, appearing to have lost all pleasure in life. Some refer to it as feeling "empty" inside.

Physical symptoms

The impact of depression can affect the body as well as the mind, and it is not uncommon for people with depression to have physical symptoms. They may suffer from unexplained aches and pains such as sore muscles, headaches, joint, and back pain. Those with depression may experience digestive problems, fatigue, exhaustion, and changes in sleep patterns. There may be weight loss or gain from changes in appetite. Slowed speech and unusual muscle movements have also been linked to depression.

Behavioral symptoms

A person experiencing clinical depression may behave differently than they usually do. They may become disinterested in people, events, and things going on around them, even in situations that they used to enjoy. Their long-held beliefs and habits may suddenly change as well. They may have trouble both falling asleep and staying asleep, and may not enjoy a good quality sleep for prolonged periods. Some other behavioral symptoms include a loss of sexual desire or interest, engaging in reckless behavior that is uncharacteristic, and increased alcohol or drug use.

Cognitive symptoms

Depression can also impact a person's ability to think clearly and to concentrate. It can impede memory, and those who suffer may have a hard time focusing, recalling events, and remembering details, as well as making decisions and choices independently. An individual with depression may experience thoughts and feelings suggesting they are worthless. They may be hyper-focused on negative thoughts and may ignore any positivity around them. These negative patterns can get out of control and lead to suicidal thoughts.

If you are someone who suffers from depression, it's important to understand that you are not alone. Depression is a widespread medical condition. Approximately 7% of Americans will experience depression at some point in a given year. Children are not immune to this disorder either. In fact, teenagers of 15 years or older are currently being diagnosed with depression at the same rate as adults in this country. Depression is one of the leading causes of disability worldwide.

Causes of Depression

There are no definitive answers for what causes depression, as it is not a simple condition with one known cause. It could be that some people are more susceptible to suffering from depressive symptoms than others. It is well accepted though that depression is often caused by a combination of different factors. We'll discuss some of the current theories on the causes of depression in the following sections.

Chemical imbalance

For some, depression is believed to be caused by an imbalance in key chemicals in the brain that is involved in mood regulation. We all have chemicals called neurotransmitters in our brain. These chemicals help different parts of our brain to communicate with each other. Some are tied tightly to our feelings of happiness and pleasure. If there is a problem with the number of neurotransmitters that we produce naturally, we may end up suffering from some symptoms of depression. The two neurotransmitters that are linked to depression are serotonin and norepinephrine.

Research suggests that the hormone cortisol, which is secreted in high doses during periods of stress, may affect some of our neurotransmitters and contribute to the development of depression. Sleep disturbances, loss of appetite, loss of sexual interest, physical exhaustion, and confusion may all be related to this biochemical imbalance.

Genetic factors

Studies show that there also may be a genetic component to the development of depression, as those with a family history of depression are more likely to receive a diagnosis themselves. The exact genes involved are not yet known. Statistically, women are nearly twice as likely as men to experience clinical depression.

Physical factors

Physical health problems can be a leading cause of depression —possibly because the mind and body are closely linked. If you are experiencing a chronic, debilitating physical health problem, this can have a negative impact on your mental health. Symptoms may develop as a result of the stress you are under while living with the illness, as well as the changes that may occur to your lifestyle and abilities. Additionally, some physical illnesses carry aspects that can lead directly to depressive symptoms.

Environmental factors

Depression can be triggered by difficult events or experiences such as a traumatic past, increased demands, or stress at work. It can occur following an extremely upsetting situation, such as the death of a loved one, a divorce, or a major financial setback. After we face a loss or traumatic event, we may experience interrupted sleep, poor appetite, and a loss of pleasure or interest in daily activities. Some of these feelings are normal if they're short-term, but may lead to a diagnosis of depression if they continue for more than a few weeks.

Periods of interpersonal conflict as well as social isolation can also have a big impact on our mental health, putting someone at a greater risk of becoming depressed.

Psychological factors

We are all affected by events in our lives differently. Our thoughts and feelings contribute to how we will experience life and how we will overcome obstacles. The way we think about the world around us can be something that is ingrained from childhood. If we're prone to viewing things in a negative way, we're more likely to experience negative emotions.

Depression, of course, has also been linked to faulty thinking patterns. Those who are depressed often magnify how bad things are in their life and frequently get stuck in negative thought processes. This can cause negativity to flourish and extend the length of the depressive episode.

Hormonal factors

It's been documented that changes in hormone production or functioning can lead to symptoms of depression in women. Women going through major changes in their hormone states, such as menopause, childbirth, thyroid problems, or similar changes, may be much more prone to depressive symptoms.

Postpartum depression will be discussed in more detail later in this chapter.

Substance abuse

Drugs and alcohol abuse has also been associated with the onset of depressive disorders. Alcohol has long been known as a depressant, especially when consumed at high levels. Excessive drug use may also lead to symptoms of depression, and this includes both recreational and prescription drug use.

Links have been shown between depression and anticonvulsants, stimulants, different types of steroids, as well as beta-blockers

Types of Depression

The word "depression" is an umbrella term for many different types of depressive mental illness. The following are several different types of depression as identified by the DSM-V (Diagnostic and Statistical Manual of Mental Disorders).

Major depression

Major depression is the most common diagnosis under this umbrella. It is characterized by a fairly lengthy period (at least two weeks) during which a person suffers from at least five or more depressive symptoms. Symptoms may include feeling sad or hopeless, inability to focus, and various changes in behavior. An episode of major depression can be quite disabling and may interfere with the individual's ability to sleep, work, socialize, and eat like a normal person. An individual may suffer from major depression a few times in their life, and episodes may occur after facing times of increased stress, loss, or breakdown in a relationship.

Persistent depressive disorder (PDD)

If an individual has depression that lasts consistently for two years or more, it is called persistent depressive disorder. This term describes a low grade, persistent chronic depression. Symptoms displayed by individuals with persistent depressive disorder may be less severe than major depression. Those diagnosed may be able to function adequately but are not at their optimum health. PDD may manifest itself as ongoing sadness, irritability, trouble concentrating, changes in habits, and inability to enjoy life.

Bipolar disorder

Bipolar disorder typically manifests itself as a shifting mood cycle. Bipolar involves severe or extreme highs followed by periods of crushing, depressive lows. This diagnosis used to be referred to as manic-depressive disorder. Symptoms generally alternate between mania and depression. When manic, a person may present themselves as very high energy and full of excitement. They may display poor judgment and a lack of inhibition. The low phase in bipolar disorder will have symptoms similar to major depression. Someone with this diagnosis may cycle through these two extremes a few times in a year or even more often than this. There are also different subtypes within this diagnosis, but for the purposes of this book, we will not get into these.

Seasonal depression/Seasonal affective disorder (SAD)

Seasonal affective disorder is represented by a period of major depression that most often happens during cold, grey, winter months. The illness seems to occur in association with lack of natural sunlight. Symptoms of SAD may include an increase in anxiety, irritability, fatigue, and weight gain. These symptoms may range from mild to severe and tend to alleviate naturally with the arrival of spring and summer months.

Peripartum (postpartum) depression

Women who experience symptoms of major depression in the weeks and months following childbirth may be diagnosed with peripartum depression. Following the birth of a child, a woman's hormone levels generally shift quite profoundly. These changing hormone levels can sometimes result in the new mother feeling depressed. Postpartum depression is characterized by the following symptoms: feelings of extreme sadness, anxiety, exhaustion, loneliness, hopelessness, fear of harming the baby, feelings of disconnect from the baby, and even suicidal thoughts.

Psychotic depression

Sometimes people can suffer such severe symptoms of major depression that they can become out of touch with reality. Their depression may be paired with additional symptoms indicative of psychosis. These individuals may experience hallucinations such as hearing voices and commands, delusions in the form of intense feelings of failure or of committing a terrible sin, as well as symptoms of general paranoia.

Situational depression

Although it is not an official term in the field of psychiatry, situational depression is a common type of depression. It can occur when you are having trouble managing a stressful situation or event in your life. Situational depression may be triggered by loss, the death of a loved one, or unexpected trauma. Some refer to this type of depression as "stress response syndrome" or "adjustment disorder." Generally, situational depression will resolve on its own without the need for medication. However, psychotherapy can help speed up this process. Symptoms may include excessive sadness, anxiety, and worry, or agitation. It is important to know that, if these symptoms do not go away in a reasonable amount of time, they may be warning signs of major depression.

Atypical depression

Atypical depression is another subset under the umbrella of depression. It generally manifests itself a little differently than the persistent sadness that is associated with typical depression. Atypical depression describes a pattern of symptoms such as increased appetite and weight gain, increased sleepiness and fatigue, reactive mood swings, and oversensitivity to rejection. With atypical depression, sometimes, a positive event can temporarily improve the individual's mood and alleviate the symptoms.

Chapter 11: Treating Depression

If you are suffering from depression, it's important to know when and where to get help. Try to see a doctor as soon as your symptoms become apparent. Many people are ashamed or afraid to ask for help as there is often a negative stigma attached to mental illness; however, by delaying treatment, your symptoms may last longer.

Some people may not seek treatment for their depression because they believe that the issue will resolve itself. Unfortunately, this is often not the case. Generally, depression does not go away without some kind of intervention. However, it is a treatable condition. If properly managed, total recovery is possible.

Your health care provider may recommend medication to treat your depression. Some medications may offer quick and effective relief for your symptoms. One caution about anti-depressants is that, although they're effective for some, they can often lead to relapse or have adverse side effects. For these reasons, many people decide against taking medications to alleviate their symptoms. It is a decision that you can make with your doctor. If you do decide to try anti-depressant medications, be aware that they can be more effective and provide longer-lasting relief when combined with psychotherapy.

Cognitive behavioral therapy can effectively treat all the types of depression discussed previously, whether in collaboration with a trained therapist or through self-directed therapy. The majority of people suffering from depression will see an improvement from participating in CBT. Oftentimes, considerable relief from symptoms can be felt within a few months of treatment.

Chapter 12: Fighting Depression with CBT

Cognitive behavioral therapy has been shown to be an excellent treatment for depression, both on its own and in conjunction with medication as prescribed by your doctor. Research has shown CBT to be very effective in treating the symptoms of depression. In fact, 80% of people with mild, moderate, or severe depression improve by following this type of psychotherapy. By focusing on the faulty, negative thinking patterns associated with depression, this type of treatment can help decrease the impact of the symptoms that depression can have on your life. Learning and practicing the skills introduced in CBT can improve the overall quality of your life.

CBT works to combat depression through the following:

- Learning what is going on in your mind and body
- Breaking down what is making you feel sad or overwhelmed
- Focusing on the thoughts, actions, and feelings that make you depressed
- Making problems more manageable
- Correcting the misinterpretations that you may have

- Controlling the negative thoughts that lead to loss of interest and feelings of worthlessness
- Helping you learn how to accept loss, disappointment, and failure and how to not blow these experiences out of proportion or dwell on them for too long
- Learning new strategies to combat sadness and hopelessness
- Learning techniques for problem-solving

Identifying and Breaking Negative Thought Patterns

Negative thinking can prolong symptoms of depression and slow your recovery. The more negative thoughts you have, the more likely you are to stay depressed. People with depression often do not allow themselves to feel positive emotions. Instead, they tend to suppress them and allow negative thoughts and feelings to take over. Even if we have positive experiences when we are clinically depressed, we can fall into the trap of telling ourselves that this happiness is not real or that we do not deserve to experience it.

The other issue is that, when suffering from depression, we tend to act on constant negative thoughts. Our feelings can be revealed in our behavior, which in turn, leads to more negative thinking, starting a downward spiral. The negative cycle can very easily cause us to feel worse about ourselves and our situation.

Before long, we can talk ourselves into feeling like a failure every time things don't go our way. Automatic thoughts start to take over the brain, causing us to dwell on unrealistic thoughts like "I always fail at everything," "I am a terrible mother," or "I can't do anything right."

Negative thought patterns can lead to feelings of hopelessness, fear, sadness, and other symptoms of anxiety and depression. The emotions attached to these thoughts may lead you to pull away from other people and become further trapped in your negative cycle.

The good news is that there are key strategies you can implement that can help counteract the symptoms associated with depression. One of the main focuses of cognitive behavioral therapy is changing the automatic negative thoughts that influence your behavior in order to break the cycle of depression. Through the CBT process, you are taught to examine and change negative thoughts to improve your emotions. You learn how to avoid giving the unhelpful thoughts in your head so much power. You also learn to acknowledge them to redirect your attention to more positive things.

The following are specific techniques that can help you identify and break negative thought patterns.

Cognitive restructuring

Cognitive restructuring is one of the main techniques focused on cognitive behavioral therapy. In this section, we will examine how this strategy can be used to combat symptoms of depression. The three main steps in cognitive restructuring are as follows.

1. Identify problematic thinking.

Use journaling techniques, as previously discussed, to identify your problematic thought patterns. Write down all of the things that are bothering you in the present. The main goal of this step is to become more aware of your negative and unrealistic thoughts. By examining them, you can get to the root of your feelings of depression. What is causing you to get caught up in feelings of hopelessness or the belief that things will never get better? What are some of the thoughts you experience when you are feeling discouraged? What do you feel helpless about? Describe the emotions and their possible causes as well as you can.

2. Write out statements to counteract your negative thoughts.

This is where you will put the thoughts and ideas that are contributing to your depression on the witness stand and come up with evidence against them. You may be surprised at how quickly you come up with counter-evidence and support against your thinking. The goal of this exercise is to develop more balanced thinking and see the positive aspects of a situation. During this step, ask yourself questions like:

- What evidence is there that this thought is true?
- Is there another way of looking at this particular situation?
- What positive perspective can I take?
- Am I only paying attention to the negative side of the event or occurrence?
- What would I tell someone else who was thinking this?

3. Find opportunities to think positive thoughts.

The next step in cognitive restructuring for depression is to come up with a positive statement to counteract each negative thought that you've written down. The idea here is not to come up with something too broad or general, as this might be far from the truth. Choose a more realistic view. Use statements that start with "I" and use the present tense. Make these statements fairly broad, practical, and about something specific to you. Carry these statements with you and read them often or memorize them. Repeat them throughout the day, especially when you're experiencing negative thoughts. Use the positive thoughts to silence the negative ones. In time, you will start to believe them, and they will soon become your automatic thoughts.

A similar technique is to finish each day by remembering and visualizing the best parts. Keep a journal of positive thoughts, sometimes called a "gratitude journal." Share these positive thoughts with other people, whether in person or online. Consciously work on forming new associations and patterns in your brain. Appreciate the beauty, happiness, and positivity around you. Get out of your unhelpful thinking habits. Use positive statements to develop a new attitude about yourself and your life.

Actively search out the good in every situation. When you are at a social event, repeat positive comments in your head. Routinely share your positive thoughts and experiences with others. In following these steps, you are retraining your brain to look for the positive in every situation automatically.

Avoid Exaggerated or Catastrophic Thinking

Do your best to avoid thought patterns that cause you to view things negatively. Such thinking can keep you from making changes or improvements. Some examples of unhealthy thought processes may include: catastrophizing, which involves focusing on the worst imagined situation or outcome, "all or nothing" thinking, which is characterized by black and white thinking, and overgeneralizations.

You may also find yourself jumping to conclusions before you have any real evidence, or using a filter to ignore positive events and suppress positive emotions. You may assume that passionate feelings or favorable situations will not last because you do not deserve them. Another type of exaggerated thinking is when you hold yourself to a strict list of what you're allowed or not allowed to do, and become hard on yourself when you break these self-imposed rules.

This type of negative thinking and feeling can fuel symptoms of depression. Using the strategies discussed above, restructure automatic negative thoughts if they do occur. Concentrate on more realistic ways of thinking about a situation.

Try to Focus on the Present

When dealing with symptoms of depression, it is important to stop ruminating on what happened in the past. Do not obsess over events or negative situations you had to deal with. You may get stuck reliving the feelings associated with these events, which can prolong episodes of depression. Thinking about what you should have done or said is not a helpful way to get past difficult situations. It is okay to reflect on what happened and record the details of an event, but you really should just focus on the how and why.

Accept the decisions you made at the time and the actions that you took. Do not take this too far and ruminate over it. On the same token, it is not good for your mental health to worry too much about the future either. Depressed individuals tend to assume they know what will happen in the future, and they usually put quite a negative spin on this. This can lead to further feelings of depression or some symptoms of anxiety. It is always a good idea to try to stay in the present as much as possible. This should make you less likely to get stuck in a situation, to exaggerate it, or to blow things out of proportion.

Practice Mindfulness

Mindfulness means being fully present in the moment, and focusing on the here and now. It involves being purposeful, non-judgmental, and practical. Mindfulness includes being aware of thoughts, feelings, sights, sounds and smells that you might not usually pay attention to. It also means slowing down your thinking, staying out of the past and the future, and concentrating on the present. Mindfulness is a skill that can be trained with practice.

Mindfulness can be very helpful in improving symptoms of depression. Learning to control your focus or attention helps you become aware of your negative thoughts and recognize their false depiction of reality. It helps decrease negative thought processes by teaching you to pay no attention to them. Mindfulness can also improve issues with concentration and focus. Practicing it regularly can help you recognize positive things and bring about new pleasures. When negative thoughts intrude, you'll learn to ignore them and continue bringing attention to more pleasant thoughts.

Accept What You Cannot Change

CBT is about learning that we cannot control everything in life and that disappointments are inevitable. But how we accept disappointment and failure can be the key to our mental health. When we are depressed, it is far too easy to take disappointments personally and develop a feeling that we have lost control of our lives.

Remembering that some things are always going to be out of our control can help us ease the pressure we sometimes place on ourselves. When something unexpected happens, do not get caught up in it or let destructive thoughts take over. Instead, focus on how you will address the situation or obstacle. It is always a good strategy to write down what happened and what your immediate response was. Think about what you learned from the situation, how you will approach it, or how you will handle it differently in the future. This is how you will begin to accept what you cannot change and learn to tolerate things that are out of your control.

A Short message from the Author:

Hey, are you enjoying the book?

I'd love to hear your thoughts!

Many readers do not know how hard reviews are to come by, and how much they help an author.

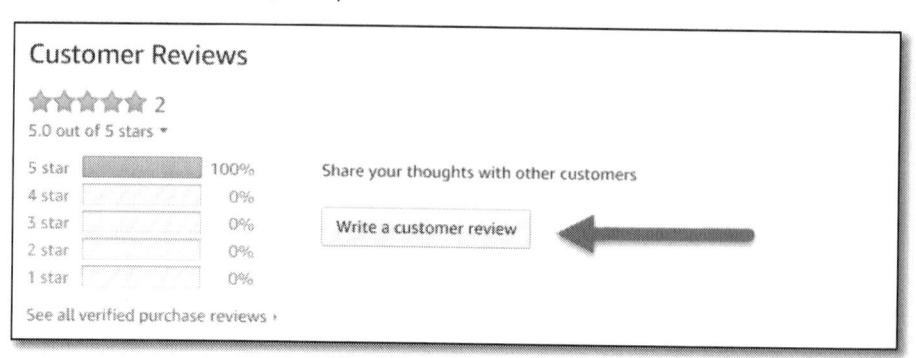

I would be incredibly grateful if you could take just 60 seconds to write a brief review on Amazon, even if it's just a few sentences!

>> Please visit **www.ReviewCBT.MindPerfection.org**

Thank you for taking the time to share your thoughts!

Your review will genuinely make a difference for me and help gain exposure for my work.

Chapter 13: Understanding Anxiety

Everyone feels anxious from time to time. For example, we may feel nervous about an upcoming event or an important decision to make. It may feel like our heart is fluttering or our stomach is upset. These feelings are a normal and natural response to a stressful situation, and can actually be beneficial when it comes to motivating us. These usually pass as quickly as they come and do not return for some time.

Anxiety disorders are different though. For some people, symptoms like these may become chronic, returning over time making the individual feel very uncomfortable and helpless. Unexplained anxiety can have an impact on normal life, including how we think, feel, and behave. Worry and fear may become overwhelming.

Anxiety disorders are the most common mental illnesses diagnosed in America. Those with the diagnosis may find themselves living with compulsive, almost constant state of worry. They may have irrational fears and thoughts, experience a variety of physical symptoms, and struggle with social interactions and relationships. Anxiety is characterized by feelings of intense discomfort, which motivates us to avoid situations that makes us uncomfortable.

No one is quite clear on what exactly causes anxiety disorders, but most agree that anxiety often stems from a complex combination of factors. One theory is that anxiety is the result of changes in our brain, like faulty circuits that control fear and other emotions. The symptoms of anxiety can also be triggered by emotional issues linked to events from the past. Similar to depression, there is some evidence that anxiety is caused by faulty and negative thought processes, such as the belief that everything has to be perfect.

The symptoms of anxiety may be triggered by a specific event, stressful experience, or social and environmental cues. They can be the result of an extreme reaction to something anxiety-provoking. Like depression, there may be a genetic component to anxiety. This disorder can affect anyone at any age, and sometimes, there doesn't seem to be a reason for its onset. There are many cases of other mental health issues leading to symptoms of anxiety.

Symptoms of Anxiety

Although there are several different types of anxiety disorders, all share the same general symptoms. Some of the main symptoms of anxiety are as follows.

Emotional symptoms

Anxiety is generally marked by feelings of panic, worry, fear, and uneasiness. There is often a general feeling of restlessness and not being able to relax. Those experiencing an anxiety attack may become quite fidgety and unable to stay still. Individuals with anxiety may become fixated on the future and often worry about things that "could" happen. It is also common for these individuals to trigger an anxiety attack just by thinking about it or talking about anxiety.

Cognitive symptoms

Individuals experiencing anxiety may deal with confusion, poor concentration, and memory deficits. Anxiety can be overwhelming, and the constant worry that is involved can make it difficult for people to concentrate. They may have difficulty making decisions and can feel stressed out or overwhelmed when given too many options.

Behavioral symptoms

Those suffering from anxiety may have a hard time falling and staying asleep. In fact, insomnia is one of the most common complaints among those with the diagnosis. Racing thoughts and extreme feelings often keep those experiencing anxiety awake through the night. Insomnia may lead to increased tiredness or exhaustion during waking hours. The symptoms of anxiety may also affect an individual's social behavior, causing them to retreat and become anti-social.

Physical symptoms

Many individuals experiencing anxiety also report a number of physical symptoms. These may include dry mouth, muscle tension, and cold, numb, or tingling hands or feet. During an anxiety attack, people may become sweaty or physically weak. When you are anxious, you may clench your muscles, which can lead to muscle tension or various aches and pains in the body.

Often, during periods of anxiety, people will experience shortness of breath, heart palpitations, and dizziness. People often report having a hard time breathing if they are feeling very anxious, especially during a panic attack. Some may deal with nausea and dizziness as well as.

Types of Anxiety

"Anxiety" is actually an umbrella term that represents a group of mental illnesses. The exact diagnosis may depend on the symptoms and the severity of the anxiety that an individual is experiencing. Anxiety disorders share the anticipation of a future threat but may differ in the types of unhealthy thoughts associated with them.

Phobias

Phobias are intense feelings of fear of a specific situation or event. The fear associated with a phobia goes beyond what is appropriate and may bring about disaster thinking and avoidance of ordinary situations. People experience severe dread when faced with the object or occasion of their phobia. They may live in constant fear of coming into contact with the specific situation and may be unable to control the fear, even knowing that it is irrational.

For some people who have severe phobias, the mere idea of the object they fear can cause stress or anxiety and leave a negative impact on their life. They may experience terror just thinking about the object of their fear. With phobias, people often change their daily habits in order to avoid the feared object. They may avoid situations that cause them to worry at all costs, restricting their habitual routines.

Panic disorder

A panic attack is a feeling of sudden and intense fear, coupled with both emotional and physical symptoms. Panic attacks usually only last for a short period, generally peaking near the ten-minute mark. Those suffering from a panic attack will often experience several physical symptoms as well, such as tingling sensations, shortness of breath, chest pain, heart palpitations, and sweating. They may feel like they are choking, having a heart attack, or feel lightheaded and dizzy. These physical sensations are typically coupled with emotional symptoms as well, such as a terrible feeling of doom, health anxiety, and intense worrying that something is very wrong.

Panic disorder can be extremely debilitating. It involves repeated and unexpected panic attacks that seem to happen with very little warning. Those with panic disorder also have to deal with the fear of panic attacks. This constant worry may cause them to change their lifestyle to avoid triggering an episode of panic.

Generalized anxiety disorder

Generalized anxiety disorder (GAD) is the most common type of anxiety. It is characterized by excessive and unrealistic worry for a period of six months or longer. Individuals with GAD may feel on edge, anxious, stressed, worried, or tense with little reason or explanation. The worry they experience is not realistic in relation to the actual situation that they are facing. Essentially, worries and concerns are blown far out of proportion. GAD is an ongoing, persistent state of mental and physical tension, often with little or no break. It can have a crippling effect on someone's life and lead to the individual affected to avoid many everyday situations in order to escape the inevitable feelings of anxiety.

Social anxiety disorder

Social anxiety disorder is also referred to as social phobia. This disorder involves an irrational fear of social situations, which goes far beyond feeling shy or discomfort in certain situations. Those with social anxiety disorder experience extreme and constant fear of being embarrassed, judged, or evaluated negatively by others in social situations and therefore, begin to avoid all potential sources of this fear. They have an irrational fear of doing something stupid or embarrassing and become extremely self-consciousness about everyday social situations.

Social anxiety disorder causes extreme shyness or awkwardness around others and can be intense enough that it causes noticeable fear and anxiety. Public situations may be very uncomfortable and distressing. As a result, people with social anxiety disorder often avoid social situations, even healthy socialization opportunities. Social anxiety disorder can lead to difficulties in coping with everyday life and generally leads those affected to display a variety of avoidance behaviors.

Agoraphobia

Agoraphobia is a type of phobia that involves fear of going out in public and unfamiliar places. It is essentially the fear of having a panic attack or experiencing feelings of anxiety in public and having no escape or solution. Agoraphobia is characterized by the fear of socializing, losing control in public, preoccupation with own safety, and irrational tension during typical situations. It can be unbearable, and many people with agoraphobia avoid public places or even avoid leaving their homes. This fear may persist for a long period and prevent the individual from living a normal, enjoyable life.

Treating Anxiety

With proper treatment, many people with anxiety disorders can learn how to manage their disruptive emotions better and get back to living a normal life. Treatment can help reduce unhelpful coping strategies and teach patients healthier behaviors. The most effective type of treatment will depend largely on the specific type of anxiety disorder that you have and the symptoms that you are experiencing. Your therapy should be tailored to these specific details. Regardless of the diagnosis, there are many effective strategies for eliminating some, if not all, of your symptoms.

Some health care providers may prescribe medication to treat your anxiety disorder. Additionally, lifestyle changes such as an increase in physical activity and a decrease in life stressors may be recommended. Some people who suffer from anxiety disorders may also find support groups to be of some help.

The most effective type of treatment for anxiety disorders is cognitive behavioral therapy. CBT can help you identify problematic patterns of thinking that increase your feelings of anxiety. It is often the first treatment recommended for mild or moderate anxiety.

Chapter 14: CBT for Treating Anxiety

Many different types of therapy can be used to treat anxiety disorders. The most widely used approach at this time is cognitive behavioral therapy. CBT is effective in the treatment of GAD, panic disorders, phobias, social anxiety disorder, agoraphobia, and more. Studies have shown that the part of the brain that is activated during fear can be changed with the help of CBT.

The two main components of CBT, cognitive therapy and behavior therapy, work together to help address and improve how you approach the world around you. CBT helps you examine how your thoughts contribute to anxiety, and how your behaviors can both trigger and prolong the symptoms. The goal with this type of therapy is to teach you that, even though you cannot control everything around you, you can at the very least learn to control how you interpret and react to your own environment.

Some of the CBT strategies recommended for tackling issues with anxiety are similar to those suggested for depression. The process and techniques may be the same, but the goal or expected outcome may be slightly different. As a whole, CBT should help you uncover the underlying causes of your worry and fear. It should teach you how to control your anxiety levels and stop negative and worrisome thoughts. CBT will help you learn to relax, even when you are under pressure, and learn to look at stressful situations in a new way. When you are suffering from symptoms of anxiety, you will usually perceive a situation to be far more dangerous than it actually is. One of the goals of CBT is to teach you how to re-shape the way you think about the events in your life that cause your symptoms. Additionally, you will learn coping strategies, better problem-solving skills, and other tools and strategies for future use.

Cognitive Restructuring

Phobias, social anxiety, generalized anxiety disorder, SAD, panic disorder, and agoraphobia are all brought about by excessive and unrealistic worry. The goal of CBT is to identify the negative thoughts that cause these worries and help you reconfigure your faulty thinking. You will learn to break patterns of negative thinking and obtain healthier ways of responding and behaving.

As previously discussed in our discussion on depression, cognitive restructuring involves challenging negative thinking patterns that contribute to mental health issues and replacing them with more realistic thoughts. This strategy can be very helpful for combating anxiety as well. Cognitive restructuring for anxiety generally involves three main steps.

1. Identify negative thoughts

People with anxiety disorders often distort situations and perceive them to be much more threatening than they really are. People may develop an irrational fear of environments, objects, or situations, and this fear may negatively impact their lives. While engaged in CBT, you will learn how to identify the irrational fears that are leading to anxiety. The first step in this process is to write down all of the things that cause you fear, worry, or anxiety, whether these feelings are warranted or not. You can keep these thoughts in a journal or record them in a "thought log." Either way, try to describe the situations that you experience and the thoughts and feelings that result. Practice reflecting on and writing down your thoughts as they occur. Make sure you include examples of self-doubt and negative self-talk. Examples of negative thoughts might be:

- Name calling, like "loser" or "idiot"
- Negative thoughts such as "No one likes me," or "Everyone is going to laugh at me."
- Having images or thoughts that something terrible is going to happen
- Believing that you are going to mess a situation up as usual

- Wishing you could just disappear from a situation
- Expecting yourself to be perfect

As you record your thoughts in the log, make sure you also identify any possible triggers or sources of your anxiety. Capture details about which situations made you feel symptoms of anxiety and how you responded physically. Describe how this made you feel and act in as much detail as you can recall.

Pay attention to the little shifts in your emotions. When you notice yourself getting anxious or agitated, try asking yourself some of the following questions:

- What am I thinking right now?
- What is making me feel anxious?
- What exactly am I worried about?
- Do I feel like something bad is going to happen? What?
- What am I telling myself right now?

2. Challenge negative thoughts.

Once you have practiced identifying your negative thoughts and you feel comfortable in recognizing them, the next step is to evaluate and question them. You want to figure out just how realistic they are. Are they exaggerated or examples of the black-and-white thinking? Are the thoughts unhelpful and lead to negative feelings and behaviors? What are the chances that a fearful event will actually unfold? Is there a good reason to feel anxious? Just because you think something doesn't mean it's true or that it will happen.

Your anxious thoughts can make you feel that your situation is hopeless when in actuality, this is highly unlikely.
Test out the beliefs you've associated with certain situations that cause you fear and anxiety. Challenging your long-held, deep-rooted beliefs can be very difficult. It might be helpful to ask some of the following questions to get yourself started:

- Is there evidence for this thought, or am I just making assumptions?
- What is the worst that could possibly happen?
- Is it likely that this will happen?
- Has it ever happened before?
- Will this matter a week from now?
- Am I overestimating the chances of a disaster?
- Am I worrying about how things should be instead of accepting the way they are?

3. Replace negative thoughts with more realistic ones.

After successfully challenging a deep-rooted belief, the next step will be to replace it with a new, positive thought. The goal is to take these irrational thoughts that you have already identified as being the cause of your anxiety and replace them with new thoughts that are more accurate. Keep in mind that the replacement thought should be close to the truth. It should not be the exact opposite of the negative thought, or else your brain will reject it. Some situations are scary, and you do not want to ignore or deny this fact. Aim to think neutrally and realistically during stressful events as putting too positive a spin on the situation might not help.

During this stage, you will want to come up with realistic, calming statements that you can repeat to yourself when you are facing a situation that generally causes you anxiety. Write these statements down and keep them in a place where you can access them when you need to. Some examples of calming statements include:

- I am not going to let this paralyze me.
- I feel confident that I can handle any situation.
- No matter what happens, I know I'm going to be okay.
- The past is over, I can't control it, and I need to forget about it.
- My anxiety is not going to get the better of me; I will get through this.

Relaxation Strategies

Relaxation techniques can be beneficial for those dealing with anxiety. Learning how to relax your mind and body can be a helpful part of therapy as physical symptoms such as tense muscles, shallow breathing, and tightness in the chest have all been linked to stress and anxiety. Relaxation strategies can help reduce symptoms of anxiety and increase feelings of calm and overall well-being.

The goal of relaxation strategies is not to eliminate or avoid all anxiety but to make it easier to ride out an episode of anxiety should it occur. Trying to prevent anxiety from happening can be all-consuming and actually cause it to get worse.

Relaxation skills are various techniques that we can use to initiate a calming response within our mind and body. You may find that you have a preference for a specific strategy that works best for you. Here, we will discuss two of the most commonly used relaxation skills: deep breathing and progressive muscle relaxation.

Deep breathing

Deep breathing is a technique that can be done anywhere at any time. There are no tools necessary, and the exercise can be effective within a few minutes or less. It requires you to take conscious control of your breathing. By practicing deep breathing, you will learn the effect that this can have on your body's relaxation response. As you are learning this new strategy, it's best to practice at least once a day, at a time that is most convenient for you. There are many variations of this technique; the following is an example of a deep breathing exercise that you can try at home.

Find a comfortable position, either lying on your back or sitting upright with feet on the floor. Close your eyes when you are ready. Begin to focus solely on your breathing. Bring attention to your belly; your abdomen should rise and fall with each deep breath you take. Place your hand on your abdomen to feel this pattern. Take a slow deep breath in through your nose and hold it for a count of 5. Slowly exhale through your mouth for a count of 6-8. As you breathe out, gently contract your stomach muscles to ensure that all air is expelled from your lungs. Repeat this cycle for at least five deep breaths. Try to keep focusing on your breathing the whole time. If you notice your mind wandering to other things, bring it back to the present.

Progressive muscle relaxation (PMR)

Progressive muscle relaxation is an easy technique to learn and can be very effective for calming an anxious mind and body. If you practice progressive muscle relaxation regularly, you can become skilled at it within a few weeks. Try to practice 1-2 times a day at the beginning of your journey. It is generally best if you do not wait until you feel anxious to try out this technique. In fact, it is often best to learn new skills when you are calm and can focus on the steps clearly.

Progressive muscle relaxation teaches you how to relax your body and mind, and lower the overall tension in your body. It can also help reduce physical symptoms such as upset stomach, body tension, and headaches, and can help you to get a good night sleep.

Progressive muscle relaxation is a simple process, involving two main steps. When practicing PMR, the first thing you will want to do is to find a comfortable spot in a quiet environment where you can direct your attention only to your body.

Take 3-4 deep breaths as you begin this exercise; then, be sure to breathe normally throughout, trying not to hold your breath. Keep your eyes closed as it may be helpful to use visualization techniques as you progress through this exercise.

Next, you will begin to slowly and progressively tense and relax your muscles, one at a time. Do this by applying muscle tension, for about five seconds at a time to a specific part of the body. Feel the tension in the muscles that you create. Next, release the tension and notice how different your muscles feel as you relax them. The tension will flow away from you as each muscle group relaxes.

Remain in this relaxed state for about 10-15 seconds before moving on to the next muscle. As you progress through the exercise, you should feel a deep relaxation slowly spreading through your body. This should feel warm and comforting. Repeat the tension and relaxation cycle until you have relaxed your entire body. Work your way from head to foot or the other way around, or whatever works best for you. Conclude this exercise with more deep breathing.

This technique will help remind your body what it feels like when it is relaxed. The more your practice, the easier it should be for you to reach a relaxed state in the future. You will also learn to recognize what tense muscles feel like as they occur throughout the day. Be careful not to hurt yourself as you practice this technique. If any muscle should give you pain as you run through the exercise, stop immediately and move on to a different muscle group.

Some people prefer to listen to a voice that guides them through the steps of progressive muscle relaxation. You can find CDs or scripts for sale with a recording to help guide you through this exercise, or you may want to search online for short videos or apps available for download.

As with any other skill, the more deep breathing and progressive muscle relaxation strategies are practiced, the more effectively and quickly they will work to alleviate your symptoms. Other strategies that you might find helpful include listening to calming sounds or guided meditation. All have been shown to help reduce signs of stress in the mind and body. Additionally, practicing yoga or other types of gentle exercise can be an excellent way to encourage your body and brain to relax.

Exposure Therapy

When we are anxious and afraid of something, our natural instinct is to avoid the situation that's causing us fear. The problem with this approach is that, by avoiding fears, we never get the chance to conquer them fully. Moreover, this type of avoidance behavior can harm the quality of your life in the future.

Avoiding our fears only makes them stronger. Every time we avoid an anxiety-provoking situation, the anxiety will only be worse the next time around. The brain interprets this behavior in the following way "When I avoid this situation, I feel better. I will go ahead and avoid this the next time too." Avoidance is effective at reducing your anxiety and fear in the short-term. However, it prevents you from tackling your fear and learning that the things you are afraid of are not so dangerous after all.

Exposure therapy is the process of facing your fears head-on. It involves exposing yourself to the situations, environments, events, or objects that make you afraid. It is a form of CBT that is most helpful for those dealing with phobias of obsessive-compulsive disorder. When using this technique, you will start with the situations, environments, or objects that cause anxiety but are tolerable. You need to stay in the situation long enough for the anxiety to reduce. When you face your fears and realize that nothing has happened to you, your anxiety will lessen in the long run.

As you experience success with this technique, you will need to start facing situations that cause you a bit more anxiety. In exposure therapy, you will not face your biggest fear right away. Gradually working your way up to your most feared situation using a step-by-step approach is called "systematic desensitization." Systematic desensitization allows you to challenge your fears gradually. It gives you confidence as you pass through the steps successfully and teaches you skills for controlling your anxiety and panic in increasingly difficult situations..

How exposure therapy works

1. Learn relaxation techniques.

Before diving into exposure therapy to treat your anxiety, it is recommended that you first become familiar with and practice some of the relaxation tips listed above. This may include deep breathing, practicing mindfulness, or progressive relaxation exercises as discussed. It is important that you first learn how to calm your mind and body before systematically exposing yourself to things that make you anxious. You can use these strategies as you confront your fears to keep your anxiety from escalating.

2. Create a list of your fears.

As you get started with exposure therapy, you will want to begin by creating a list of about ten to twenty scary situations, environments, items, or places that cause you anxiety. For example, if you are afraid of snakes and want to overcome this fear, your list may include thinking about snakes, looking at pictures of snakes, watching videos of snakes, seeing a snake in a tank, or being in a room with someone holding a snake as your scary activities. Once you have created the list, rank the items in order from the least to most anxiety-provoking activity. This is called creating a "fear hierarchy."

3. Confront the fear in real life.

Talking about what makes you anxious or writing ideas down on a list is not enough to eliminate your fears. You also need to face them head-on in a systematic, methodical way. Remember that the goal during exposure therapy is to start with the items on your list that are moderately anxiety-provoking. While you complete the first step, you will use relaxation techniques to manage your response to this fear. The idea is that, after repeated exposures, you will feel more in control, less afraid, and able to take on more challenging situations.

Consider the example with the fear of snakes. In this situation, you will likely start with the step of thinking about snakes or looking at pictures of snakes as your first goal. You will keep at this step until you start to feel less anxious doing it. You will need to regularly practice this strategy in order for it to be effective. Repeat it several times a day. You should find that as you run through the exposure therapy, your overall anxiety will not be as strong or long-lasting. The more you practice this technique, the faster your fears will fade!

4. Work through the steps.

Reaching success with the first step of your exposure therapy should boost your confidence and help you feel good about your progress. Use this motivation to keep moving further along your steps. Remember that the goal is to stay in each scary situation until your anxiety lessens. Use your relaxation techniques to keep your anxiety under control and don't try to move too quickly through each of the items on your list. If one technique is not working particularly well to keep you relaxed, try another one. Stay present in the moment and make sure that you are feeling comfortable before beginning the exposure. If it becomes difficult to move between stages, try coming up with an in-between stage that may be easier to achieve.

5. Move on to the next difficult situation.
Use the process as discussed to tackle all of the items on your list of fears. Once you experience success the first time, you will likely find it easier to tackle new situations. Create a hierarchy for each of the events and items that cause you anxiety and use the steps in exposure therapy to tackle them. Make sure that you use the relaxation techniques as you continue exposing yourself to new fears.

Recognize Rumination

CBT strategies can help you gain the ability to recognize when you are ruminating over something. Ruminating means that you are hyper-focused on something that has already happened or fear something will happen in the future. It occurs when you are repeatedly bothered by a worrisome thought and cannot get it out of your head. Ruminating is unhelpful in many ways. It reduces your problem-solving abilities because, essentially, you get stuck on the same thoughts over and over again. It can also keep you awake at night and lead to ongoing issues with insomnia. The best thing to do if you get stuck ruminating over things in your mind is to accept that you are having these thoughts, but try not to focus too much on them. Recognize that your thoughts are probably not accurate or realistic. Use mindfulness and relaxation techniques to allow all the thoughts to pass through your mind rather than trying to block them out completely, as this can cause increased agitation or anxiety, and may make the thoughts even more intrusive.

Chapter 15: Boosting Self-Esteem

Your self-image can shift depending on many factors in your environment. When you are feeling anxious or depressed, your self-esteem may suffer greatly. It is hard to have a positive self-image if you are struggling with anxious or negative thoughts and dealing with symptoms of depression or anxiety. The negative thoughts associated with these diagnoses often impact how you feel about yourself, including your body, self-image, and self-worth. These concepts may all be distorted by the dysfunctional thoughts and feelings that you experience regularly.

The term self-esteem refers to how you think and feel about yourself. It is like the collection of thoughts that you have in your mind when you think about yourself, your skills, and your abilities. Negative thoughts, telling you that you are not good enough or that you are a failure can wreak havoc on your self-esteem. How you see yourself and the perception you have of how others see you can greatly affect your feelings of self-worth and acceptance.

Research shows that low self-esteem is underpinned by the negative thought patterns and unhealthy biases we hold against ourselves. What this means is that we are often quick to notice things that confirm the negative opinions we have of ourselves while, at the same time, ignore things that point out our positive attributes. Poor self-esteem has a lot to do with excessive self-criticism and being too hard on ourselves. It can build up over many years and have quite an impact on our lives. Low self-esteem can lead to symptoms of anxiety and depression. If we do not believe in our strengths and abilities, we will not walk confidently through life. We may shy away from risks, personal relationships, and challenges and, therefore, may not gain much acceptance; causing our lack of confidence in our abilities to be reinforced and made worse.

Low self-esteem often presents itself with some of the following behaviors:
- Apologizing for things that are not your fault
- Being unable to say no to others
- Having extreme anxiety about making mistakes
- Being overly sensitive to criticism
- Displaying social isolation, withdrawal, or avoidance from others
- Feeling helpless when faced with challenges
- Desperately trying to please other people

The good news is that you can change how you see yourself through coaching and practice. CBT has proven to be an ideal approach for tackling issues related to self-worth and self-esteem. By practicing cognitive behavioral therapy, you can begin to investigate the causes of your low self-esteem and start to understand the relationship between your thoughts, feelings, and behavior. By breaking down these components, you will learn how thinking negatively and being overly critical can damage your self-esteem.

CBT for low self-esteem is about changing your perception, interpretation, and feelings about yourself. It involves a combination of changing negative thought patterns and gaining skills in mindfulness, problem-solving, assertiveness, and behavioral activation. The goal is to help you think and act like someone confident in their abilities.

Depending on your preferences and specific needs, cognitive behavioral therapy for low self-esteem may include a combination of the following techniques:

Cognitive Restructuring

Cognitive restructuring is a method of identifying negative thought patterns and the sources of how these thoughts came to be. By practicing CBT, you learn how to restructure your perceptions into thought patterns that are more positive, helpful, and effective. Cognitive restructuring can boost self-esteem by helping you identify negative thoughts you have about yourself and the distorted assumptions you've made about how others see you. It involves a combination of changing dysfunctional self-defeating thought processes, considering more realistic ways of thinking, and learning more effective behavioral responses.

1. Identify your automatic negative thoughts.
Write these thoughts down as they occur. Concentrate on self-critical and bad thoughts that you have towards yourself. Also, record the factors, events, and situations that cause you to have these negative thoughts. Examples of negative thoughts that harm your self-esteem might be:
- Other people deserve this more than me.
- I will never be good enough to succeed.
- No one wants to hear what I have to say.
- I messed up again, just like I always do.
- None of these people even like me; they can tell that I am worthless.

2. Challenge the validity of these thoughts.
Challenge your thoughts and patterns of thinking. Are they realistic? Are you making assumptions and generalizations? Is there any real evidence behind these thoughts that you are

having? Pay attention to the difference between what actually happened to you in comparison to how you interpreted it. Remember that thoughts are just thoughts, not facts. Continue to challenge every negative, self-defeating thought.

3. Rewrite the negative thoughts into healthier, more realistic, and more balanced thoughts.

Remember your strengths and good qualities, and be more accepting towards yourself. Identify the positive things, people, events, and situations around you. Carry a log book and take note of the good or helpful things you've accomplished. Pay attention to the things that you enjoy doing and the positive interactions that you have with others. Acknowledge your strengths. Reflect on your positive qualities before you go to bed every night. Write them down and repeat them to yourself in front of a mirror.

Systematic Exposure

Systematic exposure has been discussed in some detail in the section on CBT and anxiety. Exposure, as it relates to boosting self-esteem, is based on the theory that, when we avoid situations that make us uncomfortable or anxious, it prevents us from growing, gaining new perspectives, and experiencing some of the joys in life. One of the methods for dealing with low self-esteem is to systematically expose yourself to situations that make you feel distressed, a little bit at a time. By exposing yourself to situations, events, and environments that you would otherwise avoid, you can learn that they're not as bad as you thought. With practice, your fear, awkwardness, and problematic behavior will diminish naturally. As you master more difficult situations, your confidence will skyrocket.

Mindfulness Training

Research indicates that there's a correlation between low self-esteem and mindfulness. Practicing mindfulness is a way of being more aware of the present moment. It is about observing your experiences, thoughts, emotions, feelings, and sensations as they occur. It forces you to be present and have an open and non-judgmental mind. Practicing mindfulness trains your brain to observe your feelings, thoughts, and worries without getting involved or reacting to them.

In many ways, mindfulness training can help you learn to be less judgmental towards yourself. It can also reduce second-guessing and help you focus on the more positive elements. It can remind you to enjoy the simple things in life and not to worry so much. Mindfulness has been found to improve compassion, self-acceptance, positive emotions, and overall well-being.

Problem-Solving Training

When you have ongoing issues with low self-esteem, this may lead to feelings of helplessness when you're faced with unfamiliar or difficult situations. Problem-solving training can help you take a more active role in solving the problems that you face in your day to day life. This technique discourages you from passively experiencing problems and acting or feeling like a victim. It helps you learn how to identify problems and possible solutions, and come up with a plan of achieving these solutions independently and proactively.

Behavioral Activation

Behavioral activation is a technique that can be very effective for people who suffer from low self-esteem. If you have issues with confidence and self-worth, you often avoid activities and situations that you fear you won't do well in. This can lead to isolation, increasingly poor social skills, and depression.

Behavioral activation counteracts this negative cycle by helping you re-engage with the people and activities around you. It is a method whereby you identify the situations and environments that make you feel anxious and fearful and learn strategies to cope better. You break down the circumstances and the reason behind your fear. By breaking down your fears, you can analyze and question your feelings towards these situations and get to the bottom of the negative thought processes associated with them. Eventually, you feel empowered to stop avoiding the situation and engage in your life actively.

Assertiveness Training

If you are someone with low self-esteem, you may have difficulty asserting yourself, asking for what you want, saying no to other people, or expressing your true feelings. You may hide how you feel and not speak up often. Assertiveness training is a technique for learning how to reclaim your confidence and power. It involves learning new skills for getting your needs met in ways that are both effective and socially acceptable. By gaining skills through assertiveness training, you can learn to stand up for yourself and become a strong and confident communicator.

Social Skills Training

Poor social skills often go hand in hand with low self-esteem. Social skills training can help you improve your social relationships by increasing your positive interactions with others. You can learn ways to interpret social situations that you encounter and how to react accordingly. Social skills training helps decrease negative perceptions of interactions and social behavior. At the same time, it can help increase your confidence and the quality of your social engagement.

Cognitive behavioral therapy provides an individualized program based specifically on your goals and needs. The strategies recommended and practiced during the CBT process can help change negative thinking as well as core beliefs, and work out troublesome, persistent issues. When left untreated, these issues can lead to ongoing mental struggles and relapse.

Chapter 16: Overcoming Obstacles

Deciding to approach your problems using psychotherapy is not always an easy step. In this book, we have discussed a number of strategies and techniques that can be used for dealing with mental illness and low self-esteem. Those who are willing to practice these various techniques and discover what works best for them will get the most benefit out of cognitive behavioral therapy. Like any type of treatment, you may encounter limitations and obstacles as you try to progress through the steps and exercises. Below, we will look into some common barriers that might impact your progress.

Shame

Oftentimes, shame or embarrassment can be attached to mental illness. These emotions prevent many people from seeking help for their problems. Shame can make the problem worse as it can encourage feelings of depression, anxiety, low self-esteem, and the need for avoidance. It is important to understand that mental illness can happen to anyone. Some of the feelings you're experiencing may just be a normal response to a difficult situation in your life. You are not alone. You may find that taking the first step in your therapy by educating yourself and learning about different strategies for change may do a lot to reduce your shame.

Lack of Information

If this is your first experience with therapy, you may not understand entirely how it works. You may feel unsure and uncomfortable as you begin, especially if you are working with an unfamiliar therapist or coach. Educate yourself about the therapy process that you will be undertaking. Learn what you will be doing and about the process ahead. This will help you to be as comfortable as possible when you begin.

Fear of Therapy

The fear of uncertainty can create resistance and lead to avoidance behaviors. This type of anxiety can be a normal response to trying something new. Therapy can be scary at first, especially if you have never tried it before. As you get started, you may find that you are fearful about certain things like having to analyze yourself and your behaviors. You may be afraid that the strategies will not work and you will not get better, or you may have an unconscious fear of success as well. Once you take the first step in undergoing CBT, this fear will begin to diminish. During the CBT process, you will be given an opportunity to look further into your fears and explore where they are coming from.

Trust

Individuals with mental health issues commonly struggle with trusting themselves and others. In order for therapy to be successful, you will have to get real with yourself and uncover your innermost thoughts and feelings. You will need to trust the process and your own skills and strengths. It is okay if this takes some time. Trust is something that builds gradually.

Feeling Overwhelmed

Starting therapy can be very overwhelming. You will be looking into your thoughts, feelings, and behaviors, and there may be things that you do not want to uncover. It is normal to feel overwhelmed at this time; it is part of the process. Some of the relaxation techniques that you will learn in CBT can help manage feelings of fear or overwhelm. They can help you feel grounded and more hopeful about the steps you are taking to improve your mental health.

Barriers Specific to CBT

As you progress through your therapy, there may be some obstacles specifically related to cognitive behavioral therapy that get in your way. CBT involves confronting your emotions and anxieties, and this may lead you to feel a bit uncomfortable sometimes. However, you have to be open, persistent, brave and committed to the process in order to get results.

It is important to note that CBT is not a miracle cure. Do not fall into the trap of expecting too much too soon. It may take a considerable amount of time learning, analyzing, and practicing to see results.

Difficulty identifying emotions and thoughts

Some people have difficulty recognizing when their thoughts are unhealthy or irrational. It is common for our emotions to develop first before we can get a handle on the thought associated with it. This can make it difficult to see the links and causes between thoughts and feelings. It may be helpful to ask yourself a variety of questions to overcome this difficulty. Examples may include questions like "What did I tell myself at that time?" or "What was going through my mind when that happened?" Re-enacting or envisioning the situation as it happened can also help deal with this barrier. Pause during the part that you can remember entirely and see if you can discover some additional, helpful information.

Being unable to alter thinking

Another obstacle that you may face during your CBT practice is in altering or changing your negative thought patterns. You may find that, although you understand the link between your thoughts, feelings, and actions, you have a hard time changing these patterns. Even when aware of what your negative thoughts are, you're unable to replace them with positive ones. Try not to get discouraged by this. Remember that change takes time. It is important to continue to practice the techniques that you learn and to apply them in a number of different situations.

Being unable or unwilling to take risks

Changing habits and patterns take a great deal of motivation and hard work. Generally, if you have taken the first step in starting your therapy, you'll be motivated to change some part of your life. Sometimes, this motivation is short-lived though, and your willingness to continue learning, practicing, and taking risks may fade. Do your best to stick with the treatment plan that you created for yourself and be consistent with practicing your techniques. Consistency and risk-taking will lead to faster results and alleviate your symptoms sooner.

It is important to note that cognitive behavioral therapy will not address all aspects of your life. Moreover, it may not provide a lot of support for addressing past experiences or other significant events that have an impact on your mental and physical health. CBT focuses on current problems rather than the underlying causes of your condition. CBT may not be effective for everyone. Those with more complex mental health needs may require more intense intervention and additional support from their doctor.

Sorry to interrupt again, but...

Are you enjoying this book? If so, then I'd love to hear your thoughts!

As an independent author with a tiny marketing budget, I rely on readers, like you, to leave a short review on Amazon. Even if it's just a sentence or two!

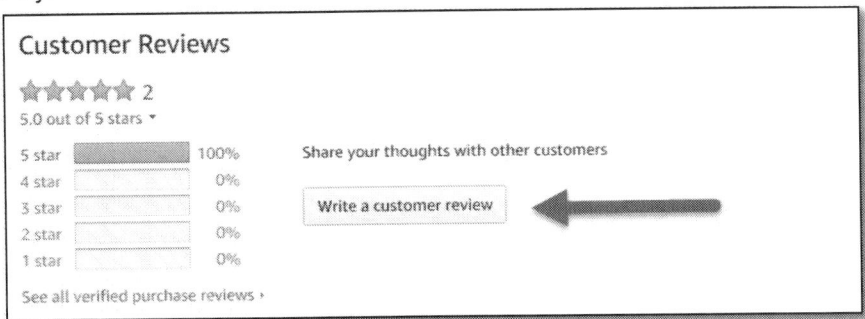

So if you enjoyed the book, please...

Visit www.ReviewCBT.MindPerfection.org and leave a brief review on Amazon.

I personally read every review, so be sure to leave me a little message.

I'd like to thank you from the bottom of my heart for purchasing this book and making it this far. And now, move on to the next page to the final chapter!

Chapter 17: Maintaining Positive Mental Health and Preventing Relapse

CBT is an evidence-based treatment and, when applied effectively, can reduce symptoms of mental illness and decrease the chances of these symptoms ever coming back. The truth is that, when symptoms of poor mental health have been reduced and you're feeling much better, it's normal to want this to last long-term. As with any medical illness though, the possibility of relapse certainly exists for those recovering from anxiety and depression. These are chronic illnesses that require a lot of work to overcome.

Relapse is the term used to describe the return to unhealthy patterns and behaviors sometime after symptoms have been reduced. It is marked by slipping back into old habits after a period of experiencing better mental health. Relapse causes you to lose some of the improvements that you have made, and if you are not careful, it may result in a complete return to old ways of thinking and behaving.

Relapse into depression or anxiety may be the result of you not using all of the new strategies that you have learned. You may still carry with you some of the old thought patterns that trigger negative thinking. During times of stress following an upsetting event, you may experience an increase in negative thoughts and thinking patterns. This may cue the depressive or anxious cycle that goes along with negative thoughts. You may find yourself returning to unhealthy routines, and may once again start having difficulties in coping with day to day activities.

To lapse briefly back into old ways is actually quite common, especially during times of stress or sadness. The occurrence of relapse may, in fact, be as high as fifty to seventy-five percent. During a relapse, it's crucial that you take steps to ensure that the relapse is only brief and symptoms do not become a full-blown issue again. Make sure that you recognize any dysfunctional thoughts as they arise, especially when you are feeling sad, overwhelmed, or stressed. It will require continued self-analysis, commitment, and management on your end to keep your illness under control.

Some studies show that relapse is less likely for those who practice cognitive behavioral therapy as a treatment for their illness. So relapses certainly do not have to take place. Here are some tips on how to prevent them.

Have a Prevention Plan

Developing a prevention plan ahead of time can help prevent relapses or limit the severity of symptoms if they do reoccur. Do not let your mental illness take over your life, but try to be mindful of the triggers and situations that are not good for you. Once you know what the warning signs look like, you can make an action plan for dealing with them quickly and effectively. Keep any helpful notes and statements with you at all times and have some go-to strategies in your mind that you can rely on quickly during times of stress.

Consider the following when creating your prevention plan:

- What future situations or events might be difficult for you?
- What can you do differently when you face them?
- Which strategies have you found to be the most helpful?
- What should you concentrate more on?
- Do you need additional support?

Keep Practicing CBT Skills

The best way to prevent a relapse into anxiety or depression is to keep practicing CBT skills regularly. You will be less likely to slide back into unhealthy patterns if you are continually working on ways to improve. Periodically using the strategies that you learn in CBT will prepare you to handle anything that comes your way.

There are a few ways you can continue practicing. For example, you might consider making a schedule of the different skills and strategies that you plan to focus on weekly. Try making a list of scenarios, situations, or objects that scare you and work on them one by one when you are ready. Work on new challenges consistently and continue to face the things that make you anxious.

Follow through on your treatment plan consistently and do not skip planned therapy sessions or discontinue until you are ready. Change things up when your growth is not progressing or your treatment plan is not working for you.

Most relapses occur during times of increased stress, loss, or change. Do your best not to dwell on negative thoughts and feelings during these times. Use your strategies for managing negative thinking patterns instead. The more you practice, the more likely you will be to prevent relapsing.

Manage Your Problems Effectively

It is also important that you continue to manage your problems effectively to avoid relapsing. Track your daily activities and progress. Keep a journal, watch for and record any shifts and increases in negative mood states. Also, think about tracking your sleep cycle and eating patterns if these have been an issue in the past.

Stay focused on how your body and mind are feeling, even during your recovery.

Know Your Triggers

In order to prevent relapse, it will be important to know your red flags and when you may be most vulnerable. Know what might trigger a possible relapse and then do what you can to prevent or minimize the influence of those triggers. Learn to recognize the who, what, when, and why's that trigger you and keep an active list of them. Track them on a calendar or schedule, look for patterns, and then anticipate them before they occur. Have a proactive plan ready for handling them. Collect information about what circumstances affect your mood and how certain situations continue to influence your behavior.

Think about your early warning signs and symptoms, and make a list of them. What does it look like when your anxiety is intensifying? How do you feel or act when your depression is becoming an issue?

Knowing what makes you vulnerable to relapse will make you less likely to have one. Be mindful of the threats to your recovery. Triggers are unique for different people, so you need to understand what makes you feel overwhelmed, and what you can do when it happens. If it helps, think back to any of your previous episodes of anxiety or depression, and determine what set you down that path in the first place.

Learn From Your Relapses

If you do happen to experience a relapse, remember that it is quite normal and take note of what you can learn from it. At the same time, try to figure out what circumstances led you to the relapse. Ask yourself questions about your thoughts and anxiety level before it happened. Was it something specific that triggered the relapse or was it a buildup of different things?

Make a plan to cope more effectively with this type of situation in the future. Remember that if you have a relapse, this does not mean that you have returned to square one. You cannot unlearn the strategies, skills, and techniques that you acquired during therapy. Recognize that you have made great improvements in the past and know how to handle your anxiety or depression at this point. You have the means to reduce your symptoms and can definitely get back on track after a relapse.

Be Kind to Yourself and Reward Yourself

Depression and anxiety can be severe illnesses to navigate. One of the things you can do to avoid a relapse successfully is to show yourself kindness and compassion on a regular basis. Make sure you take time to reward yourself for sticking to the hard work and making progress. Do things that you find enjoyable and relaxing, like reading a book or going for a walk. Pay attention to your passions, gifts, and talents, and celebrate your strengths. Pick up some good, healthy habits like exercising and eating right.

Continue to take good care of your mind and body. If you do experience a relapse, do not beat yourself up or go down the path of negative self-thinking as it will not help you get better. Be patient with yourself, learn from your relapse, and move forward. Recognize that we all make mistakes sometimes and that relapses are normal and can be overcome if you're willing to try again.

Navigating a Relapse

Sometimes, it's just not possible to prevent a relapse. But the point here is not to give up or think of it as a sign of failure. It is easy to assume that the strategies and techniques are not working. However, it is much more helpful to recognize the signs of relapse quickly and get back on track before your situation worsens. Be cognizant of early warning signs like increased irritability, sadness, and anxiety. Continue implementing what you've learned in therapy to prevent a full-blown episode. This should help to lessen the severity and duration of the relapse.

If you do fall into a relapse, be careful of what you say to yourself as this may have a negative impact on your behavior. If you start thinking that you're a failure who has undone all your hard work, it will prolong the relapse. You should also be aware of when it's time to seek additional support. Do not let your symptoms get out of control. Contact your therapist or health care provider at the first sign of symptoms and go back to treatment if necessary.

Conclusion

Thank you for making it through to the end of Cognitive Behavioral Therapy. I hope it was informative and provided you with all of the tools you need to overcome your mental illness.

Cognitive behavioral therapy is based on the concept that the meanings we attach to certain events in our lives can lead to ongoing problems. CBT follows the premise that our thoughts and feelings about a situation can strongly influence how we react to it. And thoughts, feelings, physical sensations, and actions are all interconnected this way.

CBT is a practical approach to improving your state of mind. Studies have shown that CBT has grown beyond its intended use for depression and mental illness, and is proven to be beneficial when applied to a number of different situations. CBT can be used to improve eating and sleeping habits, increase cognitive function, and improve your general well-being. Its strategies have been implemented in response to bullying at school, improving behavior in prison, and ceasing alcohol and smoking. The many different applications have great potential as a treatment plan for any number of situations.

In this book, we talked about the idea that cognitive behavioral therapy can help people understand how to improve their lives by adjusting their thinking and approach to everyday situations. In order to combat unhealthy thoughts and behaviors, the first step that we are taught is to learn to identify our problematic patterns and understand exactly how our how thoughts, feelings, and situations can contribute to maladaptive behaviors. CBT does not only involve identifying negative thought patterns but also includes learning the right strategies to use to overcome them. This process is called cognitive restructuring which is discussed several times in this book.

As you change your negative thoughts, you can readily learn to develop the skill of looking at all aspects of a situation before making conclusions. This includes recognizing the positive, negative, and neutral points. With practice, this skill should lead you to look at yourself, your situation, and your world in a more balanced and realistic way — a way that is not too negative but not too unrealistically positive either.

Learning strategies such as this one can be key to improved mental health. Other techniques discussed in this book, such as journaling, exposure therapy, relaxation techniques, and focusing on the present can also lead you to a healthier mental state. Additional strategies such as assertiveness training, social skills training, and mindfulness can also lead to many positive outcomes for those suffering from anxiety, depression, and low self-esteem.

As with any type of intervention or treatment, there may be some obstacles you need to overcome when practicing CBT to be successful. Moreover, the threat of relapse is always real when dealing with mental health issues. In this book, we discussed several strategies for both overcoming and avoiding relapse to set you on the path for positive long-term mental health.

Once again, thank you for reading Cognitive Behavioral Therapy. If you found this book useful in any way, your review is always appreciated!

If you enjoyed Cognitive Behavioral Therapy, then you'll love the next book in this series called **Emotional Intelligence Mastery.**

Now that you're equipped with the essential tools for conquering your anxiety and depression, it's time to mend relationships back to health. Learn how to start seeing the world through other people's eyes by becoming a great listener, speaking other people's language, communicating with even the most difficult individuals, and having fewer conflicts with others!

To check out this book, Visit:

www.EQ.MindPerfection.org

Printed in Great Britain
by Amazon